SpringerBriefs in Mathematics

SpringerBriefs in Mathematics showcases expositions in all areas of mathematics and applied mathematics. Manuscripts presenting new results or a single new result in a classical field, new field, or an emerging topic, applications, or bridges between new results and already published works, are encouraged. The series is intended for mathematicians and applied mathematicians.

For further volumes:
http://www.springer.com/series/10030

Daniel Alpay · Maria Elena Luna-Elizarrarás
Michael Shapiro · Daniele C. Struppa

Basics of Functional Analysis with Bicomplex Scalars, and Bicomplex Schur Analysis

 Springer

Daniel Alpay
Department of Mathematics
Ben-Gurion University of the Negev
Beer Sheva
Israel

Daniele C. Struppa
Schmid College of Science and Technology
Chapman University
Orange, CA
USA

Maria Elena Luna-Elizarrarás
Michael Shapiro
Departamento de Matemáticas
ESFM-IPN
Mexico, DF
Mexico

ISSN 2191-8198   ISSN 2191-8201 (electronic)
ISBN 978-3-319-05109-3   ISBN 978-3-319-05110-9   (eBook)
DOI 10.1007/978-3-319-05110-9
Springer Cham Heidelberg New York Dordrecht London

Library of Congress Control Number: 2014932535

Printed on acid-free paper

Springer is part of Springer Science+Business Media (www.springer.com)

Contents

Introduction

Bicomplex numbers have been studied for quite a long time, probably beginning with the work of the Italian school of Segre [1], Spampinato [2, 3], and Scorza Dragoni [4]. Their interest arose from the fact that such numbers offer a commutative alternative to the skew field of quaternions (both sets are real four-dimensional spaces), and that in many ways they generalize complex numbers more closely and more accurately than quaternions do. Of course, commutativity is gained at a price, and in this case the price is the fact that the ring of bicomplex numbers is not a field, since zero divisors arise to prevent such a possibility. Anyway, one may expect that bicomplex numbers can serve as scalars both in the theory of functions and in functional analysis, as at least a reasonable counterpart for the quaternions.

The most comprehensive study of analysis in the bicomplex setting is certainly the book of G. B. Price [5], and in recent years there has been a significant impulse to the study of the properties of those functions on the ring $\mathbb{BC}$ of bicomplex numbers, whose properties suggest a notion of bicomplex holomorphy. Rather than giving an exhaustive list of references, we refer the reader to article [6] and to the forthcoming monograph [7]. As demonstrated in these references (as well as in the more specialized papers on which those references rely), the fundamental aspects of bicomplex analysis (the analysis of bicomplex holomorphic functions) are by now fairly well understood, and it is possible to study some of the more delicate aspects of the theory of the modules of such bicomplex holomorphic functions. The history of complex analysis indicates that progress in such an arena (the study of analytic functionals, for example) cannot be achieved without a strong grasp of functional analysis. Thus, the origin of this monograph.

With the goal of providing the foundations for a rigorous study of modules of bicomplex holomorphic functions, we develop here a general theory of functional analysis with bicomplex scalars. Functional analysis in $\mathbb{BC}$ is an essentially new subject, and it seems to have had two independent starting points: the paper by Rochon and Tremblay [8] submitted and published in 2006, and the paper by Luna-Elizarrarás and Shapiro [9], originally submitted back in 2005 although published in 2009. It may be instructive to compare this with the case of functional analysis with the quaternionic scalars which dates back to the papers of Teichmuller [10] and Soukhomlinoff [11], and which is actually widely known and studied; for some recent works see, e.g., [12–17].

During the last several years the initial ideas of bicomplex functional analysis have been studied, see the papers by Gervais Lavoie, Marchildon and Rochon [18–20] published in 2010 and 2011 as well as the manuscript by Kumar, Kumar, Rochon [21]. Recently more papers have appeared, see, e.g., [22] and [23]. While there are some overlaps with our work, we present here a much more complete and exhaustive treatment of the theory, as well as many new ideas and results. In particular, we show how the general ideas that we develop for a bicomplex functional analysis can be directly employed to generalize the classical Schur analysis to the bicomplex setting.

Before we describe in detail the contents of each chapter in this monograph, we wish to point out that while the theory of bicomplex numbers is relatively young, and especially young is bicomplex functional analysis, there are many reasons why such a development should be of interest to mathematicians, engineers, and physicists.

To begin with, we will address why the study of bicomplex analysis is of general mathematical interest. Functional analysis historically began with the study of linear objects (spaces and operators) with real and complex scalars; many deep and far-reaching results have been obtained since then. The logic of the development of mathematical structures has led to the study of modules over rings, that is, to the linear objects where the "scalars" are elements of an arbitrary ring. Obviously, in this case one loses many important properties of real and complex numbers and thus the corresponding functional analysis has no chances to obtain reasonably good analogs of the basic results of the classical one. It is therefore reasonable to look for different sets of scalars, somewhat intermediate between real and complex numbers, on one hand, and an arbitrary ring, on the other. The first such interesting object is the skew field of quaternions and, indeed, functional analysis with quaternionic scalars has been able to obtain a number of basic facts. The importance of such functional analysis cannot be understated, especially as it applies to the quaternionic reformulation of quantum physics (see [24]), and to this purpose a new functional calculus, which allows the study of quaternionic operators, has been developed, see, e.g., [25]. Despite the significant successes of these theories, however, it is undeniable that the lack of commutativity of quaternionic multiplication causes many essential difficulties.

One can then consider the set of bicomplex numbers, which is, like the set of quaternions, a four-dimensional real algebra; in contrast to quaternions, however, bicomplex multiplication is commutative and possesses a number of additional properties which allow a construction of a rich theory of linear objects. The first five chapters of this monograph are a confirmation of this thesis and we believe they will become the basis of a much wider theory of functional analysis with bicomplex scalars. We should as well point out that other mathematicians, for example, the Russian mathematician A. Khrennikov, have developed a substantial amount of work on what he calls "complex hyperbolic" analysis and, which we call instead "bicomplex." We refer the interested reader to some of his most relevant articles, namely [26–29]. We also call the attention of the reader to a series of works on solutions of the complex Laplacian (functions usually referred to as complex harmonic functions), whose theory is also closely related to what we discuss in this monograph: the reader may want to review [30–33], and the references contained in those papers.

But functional analysis in the bicomplex setting is not only relevant from a mathematical point of view, but also has important applications in engineering and physics. Let us begin by making a general remark on Schur analysis (the main application discussed in our monograph). Schur analysis can be seen as the theory of self-mappings of the open unit disk (also known as Schur functions), with emphasis on applications to operator theory, linear system theory, and analytic function theory. Such theory has been considered in a number of settings besides the open unit disk, *and in each case the interaction between the setting in question and the original problems from Schur analysis has led to new insights and problems*. We mention here the case of several complex variables (see [34], where the open unit disk is replaced by the polydisk, and [35–37] for further examples), the case of Riemann compact surfaces, where the Riemann sphere is replaced by a compact Riemann surface (see [38, 39]), and the more recent developments of Schur analysis in the setting of slice hyperholomorphic functions [40, 41].

Such an interaction is developed in the present work in the setting of bicomplex numbers. As it is often (but not always) the case, engineers were precursors and a number of papers have appeared in the recent years studying digital signal processing and associated filters in the setting of bicomplex numbers (which they often called reduced biquaternions). We mention as a sample [42–47], and, for a study in the setting of hyperbolic (and multi-hyperbolic numbers), [48]. The motivation in these papers to use the bicomplex numbers is to have augmented parallelism and more efficient computations. An important question is to have efficient algorithms to multiply bicomplex numbers; see [49]. The problems considered in these papers are thus mostly of an algebraic nature, only rational functions and finite matrices are involved, and no infinite dimensional analysis is involved there. Paraunitary filters (that is, filters which take unitary values on the unit circle), with coefficients in the bicomplex numbers are studied in [42]. Our theory of realization presented in the present book should allow to make connection with, and use the results of, [50, 51] to study new aspects of these filters.

To explain the connections with our work that occur in linear system theory and digital signal processing, recall that transfer functions (or filters) of linear time-invariant dissipative systems are functions that are both analytic and contractive in the open unit disk, that is, Schur functions. The Schur algorithm associates to such a function a sequence of numbers of modulus strictly less than 1, which lead to the lattice filters in electrical engineering, see, for instance, [52, 53]. Although the aim of the developments is to have effective computations on the level of matrices, deep tools of functional analysis play an important role. We mention in particular the theory of reproducing kernel Hilbert spaces, operator models, and inverse scattering (see [54] for a survey). The theory we develop in this monograph should allow to extend the algebraic analysis to more general structures, and in particular in the setting of inverse scattering.

The above discussion considers the discrete case. When considering linear continuous systems, functions analytic and with a real positive part in the open right half-plane come into play. As appears from Zemanian's work [55, 56], the continuous case involves the theory of topological vector spaces and Schwartz kernel theorem (see, for instance, [57]). Developing the analogous theory in the bicomplex setting requires to have functional analysis on a strong basis, and our work is conducive to that.

We conclude this short detour by pointing out that just like quaternions have been used to describe a quaternionic version of quantum physics [24], so one can attempt a bicomplex version of quantum physics. There are at least two major current research efforts in this direction, one led by Rochon and his collaborators (see, e.g., [18–23]), and the other associate to the work of Khrennikov (see, e.g., [26–29]).

The work of Rochon and collaborators has been already mentioned in other parts of the introduction, and its physical context consists in the replacement of the complex field with the ring of bicomplex numbers in the formulation of quantum mechanics. In particular, the authors study the bicomplex variation of the Schrödinger equation, and the bicomplex variation of the harmonic oscillator. Whether or not the use of bicomplex scalars will provide additional physical insight is still to be determined. If the theory is to succeed, however, the kind of analysis that we present in this monograph will prove to be crucial. Similarly important will be the development of a bicomplex functional calculus. As the work in [25] and in the subsequent work of some of our collaborators [59] has demonstrated, the quaternionic functional calculus is an important instrument in the development of an operational quaternionic quantum mechanics. So, we believe it will be now important to develop an analog of the results in [38] for the case of operators on bicomplex moduli. The first step in this direction has been recently taken in [60], where the case of bounded operators on bicomplex moduli is treated.

On the other hand, Khrennikov develops a calculus of pseudo-differential operators over the ring of hyperbolic numbers, a natural subring of bicomplex numbers, consisting of numbers of the form $x + jy$ with $x, y \in \mathbb{R}$ and $j^2 = 1$. In this setting, the physical states are now represented by vectors in what the author calls Hilbert Hyperbolic Modules. Khrennikov then goes on to introduce hyperbolic-valued distributions and ultradistributions and is able to reconstruct the main points of the Paley-Wiener theory. At least from a mathematical point of view, there are enough aspects of his work that might inspire further analysis and research. But from a physical point of view, the most interesting idea is the suggestion that it may be possible to consider a hyperbolic rule of interference of quantum probabilities, rather than the usual trigonometric rule that is used customarily. By the same token, as the author points out, the formalism that he introduces is essentially a theory of hyperbolic probability waves, instead of the complex-based theory of trigonometric probability waves. Khrennikov also suggests that it is indeed possible that hyperbolic interference could even be recorded in experiments with elementary particles, though we are not aware of any such experimental result.

Let us now describe in more detail the contents of each chapter in this monograph.

Even though the basic properties of bicomplex numbers are well known and widely available, our analysis requires some more delicate discussion of the various structures that are hidden in the ring of bicomplex numbers. For this reason we use Chap. 1 to study in detail the subset $\mathbb{D}$ of hyperbolic numbers, and we are able to establish why we claim that $\mathbb{D}$ plays, for bicomplex numbers, the same role that $\mathbb{R}$ plays for complex numbers. In particular, we introduce a new partial order on $\mathbb{D}$, which has interesting connections with the Minkowksi space of special relativity, and that allows us to introduce a new hyperbolic-valued norm on the set of bicomplex numbers.

In Chap. 2, these ideas are extended to the case of matrices with bicomplex entries. Although no surprises occur here, the study of bicomplex matrices is a quite "fresh" object of research and there is not much information about them although some important facts can be found in Gervais Lavoie, Marchildon, and Rochon [20]. We present here a detailed description of some peculiarities which arise in the study of bicomplex matrices. Besides, we include into this chapter a quick survey of holomorphicity in the bicomplex setting.

In Chap. 3, we study $\mathbb{BC}$-modules, and we show their subtle nature as well as the many structures that one can impose on them. In particular, we will show that every $\mathbb{BC}$-module decomposes into the direct sum of submodules, which we call the idempotent decomposition of a $\mathbb{BC}$-module (a notion inherited from the idempotent decomposition for bicomplex numbers): such a decomposition will play a key role in the remainder of the work. We also show how to use two complex linear spaces to generate a $\mathbb{BC}$-module of which they are the idempotent representation.

The most surprising portion of the work takes place in Chap. 4, where we study inner products and norms in bicomplex modules. The matter is that depending on the kind of the scalars an inner product can take real, complex, or quaternionic values (they can be even more general) but the corresponding norm is always real-valued. We consider two kinds of norms on bicomplex modules: a real-valued norm (as one would expect) and a hyperbolic-valued norm. Interestingly enough, while both norms can be used to build the theory of normed bicomplex modules, the hyperbolic-valued norm appears to be much better compatible with the structure of $\mathbb{BC}$-modules. Note also that we apply the developed tool for a study of the ring $\mathbb{H}(\mathbb{C})$ of biquaternions (complex quaternions) seen as a $\mathbb{BC}$-module. Since the $\mathbb{H}(\mathbb{C})$-valued functions arise in a wide range of areas we believe that our analysis will be rather helpful for constructing the theory of linear space of such functions.

Chapter 5 sets the stage for the study of linear functionals on bicomplex modules. The results in this chapter are sometimes surprising. Recent works have studied spaces of analytic functionals on spaces of bicomplex holomorphic functions, for example, see [61], but so far the duality has been defined by considering only the $\mathbb{C}$-linear structure. We expect the results from this chapter to allow a new understanding of those spaces of analytic functionals and a deeper description of the duality properties.

Finally, in Chap. 6, we describe a bicomplex version of the classical Schur analysis. This is a significant application of the theory developed in the first five chapters of this monograph. Schur analysis (see [63] for a survey) studies holomorphic contractive functions in the open unit disk of $\mathbb{C}$, and has many connections and applications to interpolation problems, moment problems, and the theory of linear systems. In this final chapter of our work, we show how these same ideas can be extended in a very successful way to the case of bicomplex spaces, and we consider, in this setting, Blaschke factors, the Hardy spaces, and the notion of Schur multipliers and their realizations.

The Mexican authors have been partially supported by CONACYT projects as well as by Instituto Politécnico Nacional in the framework of COFAA and SIP programs; they are also grateful to Chapman University for the support offered in preparing this work. M. J. C. Robles-Casimiro helped with the preparation of the figures which is appreciated by the authors. Daniel Alpay wishes to thank the Earl Katz family for endowing the chair, which supported his research.

References

1. C. Segre, Le rappresentazioni reali delle forme complesse e gli enti iperalgebrici. Math. Ann. **40**, 413–467 (1892)
2. N. Spampinato, Estensione nel Campo Bicomplesso di Due Teoremi, del Levi–Civita e del Severi, per le Funzione Olomorfe di Due Variablili Complesse. I, II, Atti Reale Accad. Naz. Lincei, Rend **22**(6), 38–43, 96–102 (1935)
3. N. Spampinato, Sulla Rappresentazione delle Funzioni di Variabile Bicomplessa Totalmente Derivabili. Ann. Mat. Pura Appl. **14**(4), 305–325 (1936)
4. G. Scorza Dragoni, Sulle funzioni olomorfe di una variabile bicomplessa. Reale Accad. dItalia, Mem. Classe Sci. Nat. Fis. Mat. **5**, 597–665 (1934)
5. G.B. Price, *An Introduction to Multicomplex Spaces and Functions* (Marcel Dekker, New York, 1991)
6. M.E. Luna–Elizarrarás, M. Shapiro, D.C. Struppa, A. Vajiac, Bicomplex numbers and their elementary functions. Cubo, A Math. J. **14**(2), 61–80 (2012)
7. M.E. Luna–Elizarrarás, M. Shapiro, D.C. Struppa, A. Vajiac, Bicomplex holomorphic functions: the algebra, geometry and analysis of bicomplex numbers (book in preparation)
8. D. Rochon, S. Tremblay, Bicomplex quantum mechanics II: the Hilbert space. Adv. Appl. Cliffod algebras **16**, 135–157 (2006)
9. M.E. Luna Elizarrarás, M. Shapiro. On modules over bicomplex and hyperbolic numbers. In R.K. Kovacheva, J. Lawrynowicz, S. Marchiafava (eds.), *Applied Complex and Quaternionic Approximation* (Edizioni Nuova Cultura, Rome, 2009), pp. 76–92
10. O. Teichmüller, Operatoren im Waschsschen Raum. Journal für Mathematik Bd 174, Heft 1/2, 73–125 (1936)
11. G.A. Soukhomlinoff. Über Fortsetzung von linearen Funktionalen in linearen komplexen Räumen und linearen Quaternionräumen. Mat. Sb. (N.S.) 3, 353–358 (1938)

12. D. Alpay, M. Shapiro. Reproducing kernel quaternionic Pontryagin spaces. Integr. Eqn. Oper. Theory **50**, 431–476 (2004)

13. D. Alpay, M.E. Luna-Elizarrars, M. Shapiro. Normes des extensions quaternionique d'opérateurs réels. Comtes Rendus de l'Académie des Sciences Mathematique, Ser. I, 340, 639–643 (2005)

14. D. Alpay, M.E. Luna-Elizarrarás, M. Shapiro. On Norms of quaternionic extensions of real and complex mappings. Math. Methods Appl. Sci. **30**, 1005–1036 (2007)

15. M.E. Luna-Elizarrarás, M. Shapiro. On some properties of quaternionic inner product spaces 25th International Colloquium on Group Theoretic methods in Physics, Cocoyoc, Mexico, 2–6 August 2004. Inst. Phys. Conf. Ser. **185**, 371–376, (2005)

16. M.E. Luna–Elizarrarás, M. Shapiro. Preservation of the norms of linear operators acting on some quaternionic function spaces. Oper. Theory Adv. Appl. **157**, 205–220 (2005)

17. M.E. Luna-Elizarrarás, M. Shapiro. On some relations between real, complex and quaternionic linear spaces. In H. Begehr, F. Nicolosi (eds.), *More Progresses in Analysis* (World Scientific, Singapore, 2009), pp. 999–1008

18. R. Gervais Lavoie, L. Marchildon, D. Rochon. Infinite–dimensional bicomplex Hilbert spaces. Ann. Funct. Anal. **1**(2), 75–91 (2010)

19. R. Gervais Lavoie, L. Marchildon, D. Rochon. Hilbert space of the bicomplex quantum harmonic oscillator. AIP Conf. Proc. **1327**, 148–157 (2011)

20. R. Gervais Lavoie, L. Marchildon, D. Rochon. Finite–dimensional bicomplex Hilbert spaces. Adv. Appl. Clifford Algebras **21**(3), 561–581 (2011)

21. Rajeev Kumar, Romesh Kumar, D. Rochon. The fundamental Theorems in the framework of bicomplex topological modules, arXiv:1109.3424 (2011)

22. K.S. Charak, R. Kumar, D. Rochon. Infite dimensional bicomplex spectral decomposition theorem. Adv. Appl. Clifford Algebras **23**, 593–605 (2013)

23. K.S. Charak, R. Kumar, D. Rochon. Bicomplex Riesz–Fischer Theorem. Glob. J. Sci. Front. Res. Math. Decis. Sci.**13**–**F**(1), 67–77 (2013)

24. S.L. Adler, in *Quaternionic Quantum Mechanics and Quantum Fields* (Oxford University Press, New York, 1995)

25. F. Colombo, I. Sabadini, D.C. Struppa, in *Non-Commutative Functional Calculus: Theory and Applications of Slice Hyperholomorphic Functions*. Progress in Mathematics, Vol. 289 (Birkhäuser, Basel, 2011)

26. A. Khrennikov, Hilbert space over complex hyperbolic numbers and hyper–trigonometric interference. Infinite Dimension. Anal., Quantum Probab. Related Top. **12**(3), 469–478 (2009)

27. A. Khrennikov, Fourier analysis over hyperbolic algebra, pseudo–differential operators, and hyperbolic deformation of classical mechanics. Infinite Dimension. Anal., Quantum Probab. Related Top. **10**, 421–438 (2007)

28. A. Khrennikov, Hyperbolic quantum mechanics. Dokl. Acad. Nauk **402**(2), 170–172 (2005). English Translation: Dokl. Math. **71**(3), 363–365 (2005)

29. A. Khrennikov, Hyperbolic quantum mechanics. Adv. Appl. Clifford Algebras **13**(1), 1–9 (2003)

30. K. Fujita, Hilbert spaces related to harmonic functions. Tôhoku Math. J. **48**, 149–163 (1996)

31. K. Fujita, Topics on the Bergman kernel for some balls. More Progresses in Analysis, Proceedings of the 5th International ISAAC Congress, Catania, July 2005, pp. 125–134

32. M. Morimoto, Analytic functionals on the sphere. Translations of mathematical monographs. Am. Math. Soc. **178** (1998)

33. M. Morimoto, K. Fujita, Conical Fourier–Borel transformations for harmonic functionals on the Lie ball. *Generalizations of Complex Analysis* (Banach Center Publications, Inst. Math. Polish Acad. Sci., 37, 1996), 95–113

34. J. Agler, On the representation of certain holomorphic functions defined on a polydisk. Oper. Theory Adv. Appl. **48**, 47–66 (1990)

35. J. Ball, T. Trent, V. Vinnikov, Interpolation and commutant lifting for multipliers on reproducing kernel Hilbert spaces. In Proceedings of Conference in honor of the 60th birthday of M.A. Kaashoek, volume 122 of Operator Theory: Advances and Applications, Birkhauser, 2001, pp. 89–138
36. J. Ball, V. Bolotnikov, Nevanlinna-Pick interpolation for Schur-Agler class functions on domains with matrix polynomial defining function in C^n. New York J. Math. **11**, 247–290 (2005) (electronic)
37. J. Ball, V. Bolotnikov, A bitangential interpolation problem on the closed unit ball for multipliers of the Arveson space. Integr. Eqn. Oper. Theory **46**, 125–164 (2003)
38. V. Vinnikov, Commuting operators and function theory on a Riemann surface. In Holomorphic spaces (Berkeley, CA, 1995), pp. 445–476 (Cambridge University Press, Cambridge, 1998)
39. D. Alpay, V. Vinnikov, Finite dimensional de Branges spaces on Riemann surfaces. J. Funct. Anal. **189**(2), 283–324 (2002)
40. D. Alpay, F. Colombo, I. Sabadini, Schur functions and their realizations in the slice hyperholomorphic setting. Integr. Eqn. Oper. Theory **72**, 253–289 (2012)
41. D. Alpay, F. Colombo, I. Sabadini, Pontryagin de Branges Rovnyak spaces of slice hyperholomorphic functions. Journal d'analyse mathématique **121**, 87–125 (2013)
42. D. Alfsmann, H.G. Göckler, *EUROCON 2005: The International Conference on Computer as a Tool, 2005.* Design of hypercomplex allpass-based paraunitary filter banks applying reduced biquaternions, vol. 1 (2005), pp. 92–95
43. Xiao-Feng Gong, Zhi-Wen Liu, You-Gen Xu, Coherent source localization: bicomplex polarimetric smoothing with electromagnetic vectorsensors. IEEE Transactions on Aerospace and Electronic Systems, **47**, 2268–2285 (2011)
44. K. Okabayashi, S. Takahashi, Reference networks for simultaneous realization of complex IIR digital filters. IEEE Pacific Rim Conference on Communications, Computers and Signal Processing, 1997. 10 Years PACRIM 1987–1997—Networking the Pacific Rim, vol. 1 (1997), pp. 165–168
45. Hans-Dieter Schütte, Jörg Wenzel, Hypercomplex numbers in digital signal processing. In IEEE International Symposium on Circuits and Systems, vol. 2 (1990), pp. 1557–1560
46. H. Toyoshima, Computationally efficient implementation of hypercomplex digital filters. In Proceedings of the 1998 IEEE International Conference on Acoustics, Speech and Signal Processing, vol. 3 (1998)
47. K. Ueda, S. Takahashi, Digital filters with hypercomplex coefficients. Electronics and Communications in Japan (Part III: Fundamental Electronic Science), vol. 76(9), 85–98 (1993)
48. D. Alfsmann, H.G. Göckler, Hypercomplex bark-scale filter bank design based on allpass-phase specifications. In Proceedings of the 20th European Signal Processing Conference (EUSIPCO), 2012, pp. 519–523
49. V.S. Dimitrov, T.V. Cooklev, B.D. Donevsky, On the multiplication of reduced biquaternions and applications. Inform. Process. Lett. **43**(3), 161–164 (1992)
50. D. Alpay, I. Gohberg, Unitary rational matrix functions. In I. Gohberg (ed.), *Topics in Interpolation Theory of Rational Matrix-Valued Functions.* Operator Theory: Advances and Applications, vol. 33 (Birkhäuser Verlag, Basel, 1988), pp. 175–222
51. J. Ball, I. Gohberg, L. Rodman, Interpolation of rational matrix functions. Operator Theory: Advances and Applications, vol. 45 (Birkhäuser Verlag, Basel, 1990)
52. A.M. Bruckstein, T. Kailath, Inverse scattering for discrete transmission-line models. SIAM Rev. **29**(3), 359–389 (1987)
53. T. Kailath, A theorem of I. Schur and its impact on modern signal processing. In I. Gohberg (ed.), I. Schur methods in operator theory and signal processing. Operator Theory: Advances and Applications, vol. 18 (Birkhäuser Verlag, Basel, 1986), pp. 9–30

54. D. Alpay, in *The Schur Algorithm, Reproducing Kernel Spaces and System Theory Translated from the 1998 French Original* ed. by S. Stephen (Wilson Panoramas et Synthèses, American Mathematical Society, 2001)

55. A.H. Zemanian, in *Distribution Theory and Transform Analysis*, 2 nd edn. An introduction to generalized functions, with applications (Dover Publications Inc., New York, 1987)

56. A.H. Zemanian, in*Realizability Theory for Continuous Linear Systems* (Dover Publications, Inc., New York, 1995)

57. F. Treves, *Topological Vector Spaces, Distributions and Kernels* (Academic Press, 1967)

58. R. Gervais Lavoie, L. Marchildon, D. Rochon, Infinite–dimensional bicomplex Hilbert spaces. Ann. Funct. Anal. **1**(2), 75–91 (2010)

59. F. Colombo, I. Sabadini, The quaternionic evolution operator. Adv. Math. **227**, 1772–1805 (2011)

60. F. Colombo, I. Sabadini, D.C. Struppa, Bicomplex holomorphic functional calculus. Math. Nachtrichten, 2014. doi:10.1002/mana.201200354

61. R. Gervais Lavoie, L. Marchildon, D. Rochon, Finite–dimensional bicomplex Hilbert spaces. Adv. Appl. Clifford algebras, **21**(3), 561–581 (2011)

62. D.C. Struppa, A note on analytic functionals on the complex light cone. In G. Gentili, I. Sabadini, M. Shapiro, F. Sommen, D.C. Struppa (eds.), Advances in Hypercomplex Analysis, Springer INdAM Series 1 (2013), pp. 119–124.

63. D. Alpay, The Schur algorithm, reproducing kernel spaces and system theory translated from the 1998 French original by Stephen S. Wilson Panoramas et Synthèses. Am. Math. Soc. (2001)

Chapter 1
Bicomplex and Hyperbolic Numbers

Abstract The main properties of bicomplex and hyperbolic numbers are considered, in particular, the three conjugations on them generate the corresponding moduli of a bicomplex number which are not real valued: two of them are complex valued and one is hyperbolic valued. The notion of a positive hyperbolic number allows to introduce a partial order on the set of hyperbolic numbers which has far-reaching consequences in the theory of normed bicomplex modules.

Keywords A partial order on hyperbolic numbers · Bicomplex and hyperbolic numbers · Hyperbolic valued modulus of a bicomplex number · Positive hyperbolic numbers

1.1 Bicomplex Numbers

The ring of bicomplex numbers is the commutative ring $\mathbb{BC}$ defined as follows:

$$\mathbb{BC} := \{ Z = z_1 + z_2 \mathbf{j} \mid z_1, z_2 \in \mathbb{C}(\mathbf{i}) \},$$

where $\mathbf{i}$ and $\mathbf{j}$ are commuting imaginary units, i.e.,

$$\mathbf{i}\mathbf{j} = \mathbf{j}\mathbf{i}, \quad \mathbf{i}^2 = \mathbf{j}^2 = -1,$$

and $\mathbb{C}(\mathbf{i})$ is the set of complex numbers with the imaginary unit $\mathbf{i}$. Observe that for the particular case where $z_1 = x_1$ and $z_2 = x_2$ are real numbers, $Z = x_1 + x_2 \mathbf{j}$ is a complex number with the imaginary unit $\mathbf{j}$. Since the two imaginary units $\mathbf{i}$ and $\mathbf{j}$ coexist inside $\mathbb{BC}$, in what follows we will distinguish between the two sets of complex numbers $\mathbb{C}(\mathbf{i})$ and $\mathbb{C}(\mathbf{j})$. Note also that since $(\mathbf{i}\mathbf{j})^2 = 1$, if $z_1 = x_1$ is real and $z_2 = y_2 \mathbf{i}$ is a purely imaginary number, one has that $Z = x_1 + y_2 \mathbf{i}\mathbf{j}$ is an element of the set $\mathbb{D}$ of hyperbolic numbers, defined as

D. Alpay et al., *Basics of Functional Analysis with Bicomplex Scalars,
and Bicomplex Schur Analysis*, SpringerBriefs in Mathematics,
DOI: 10.1007/978-3-319-05110-9_1, © The Author(s) 2014

$$\mathbb{D} := \left\{ a + b\,\mathbf{k} \mid a, b \in \mathbb{R}, \ \mathbf{k}^2 = 1, \ \mathbf{k} \notin \mathbb{R} \right\}.$$

Thus the subset $\{ x_1 + y_2\,\mathbf{i}\,\mathbf{j} \mid x_1, y_2 \in \mathbb{R} \}$ in $\mathbb{BC}$ is isomorphic (as real algebra) to $\mathbb{D}$ and can be identified with it.

We will soon see that in addition to containing two copies of the complex numbers, the whole set $\mathbb{BC}$ has many similarities with the set of complex numbers although, of course, many differences, sometimes quite striking, also arise.

1.2 Conjugations and Moduli

Any bicomplex number can be written in six different ways, which are relevant for what follows:

$$
\begin{aligned}
Z &= (x_1 + \mathbf{i}\,y_1) + (x_2 + \mathbf{i}\,y_2)\mathbf{j} &=&: z_1 + z_2\,\mathbf{j} \\
&= (x_1 + x_2\,\mathbf{j}) + (y_1 + y_2\,\mathbf{j})\,\mathbf{i} &=&: \eta_1 + \eta_2\,\mathbf{i} \\
&= (x_1 + \mathbf{i}\mathbf{j}\,y_2) + \mathbf{j}(x_2 - \mathbf{i}\mathbf{j}\,y_1) &=&: \mathfrak{z}_1 + \mathbf{j}\,\mathfrak{z}_2 \\
&= (x_1 + \mathbf{i}\mathbf{j}\,y_2) + \mathbf{i}(y_1 - \mathbf{i}\mathbf{j}\,x_2) &=&: \mathfrak{r}_1 + \mathbf{i}\,\mathfrak{r}_2 \\
&= (x_1 + \mathbf{i}\,y_1) + \mathbf{i}\mathbf{j}(y_2 - \mathbf{i}\,x_2) &=&: \alpha_1 + \mathbf{k}\,\alpha_2 \\
&= (x_1 + \mathbf{j}\,x_2) + \mathbf{i}\mathbf{j}(y_2 - \mathbf{j}\,y_1) &=&: \nu_1 + \mathbf{k}\,\nu_2 \,,
\end{aligned}
\tag{1.1}
$$

where $z_1, z_2, \alpha_1, \alpha_2 \in \mathbb{C}(\mathbf{i})$, $\ \eta_1\,\eta_2, \nu_1, \nu_2 \in \mathbb{C}(\mathbf{j})$ and $\mathfrak{z}_1, \mathfrak{z}_2, \mathfrak{r}_1, \mathfrak{r}_2 \in \mathbb{D}$. We present all these representations of bicomplex numbers not only for completeness but also because all of them manifest themselves in the study of the structure of the modules with bicomplex scalars. We will later introduce one more representation, known as the idempotent representation of a bicomplex number.

Since $\mathbb{BC}$ contains two imaginary units whose square is -1, and one hyperbolic unit whose square is 1, we can consider three conjugations for bicomplex numbers in analogy with the usual complex conjugation:

(I) $\overline{Z} := \overline{z}_1 + \overline{z}_2\,\mathbf{j}$ (the bar-conjugation);

(I) $Z^\dagger := z_1 - z_2\,\mathbf{j}$ (the †–conjugation);

(III) $Z^* := \left(\overline{Z} \right)^\dagger = \overline{\left(Z^\dagger \right)} = \overline{z}_1 - \overline{z}_2\,\mathbf{j}$ (the $*$–conjugation),

where $\overline{z}_1, \overline{z}_2$ denote the usual complex conjugates to z_1, z_2 in $\mathbb{C}(\mathbf{i})$.

Let us see how these conjugations act on the complex numbers in $\mathbb{C}(\mathbf{i})$ and in $\mathbb{C}(\mathbf{j})$ and on the hyperbolic numbers in $\mathbb{D}$. If $Z = z_1 \in \mathbb{C}(\mathbf{i})$, i.e., $z_2 = 0$, then $Z = z_1 = x_1 + \mathbf{i}y_1$ and one has:

$$\overline{Z} = \overline{z}_1 = x_1 - \mathbf{i}y_1 = z_1^* = Z^*, \qquad Z^\dagger = z_1^\dagger = z_1,$$

that is, both the bar-conjugation and the $*$-conjugation, restricted to $\mathbb{C}(\mathbf{i})$, coincide with the usual complex conjugation there, and they both fix all elements of $\mathbb{C}(\mathbf{i})$.

If $Z = \eta_1$ belongs to $\mathbb{C}(\mathbf{j})$, that is, $\eta_1 = x_1 + x_2\,\mathbf{j}$, then one has:

$$\overline{\eta_1} = \eta_1, \qquad \eta_1^* = x_1 - x_2\mathbf{j} = \eta_1^\dagger,$$

that is, both the $*$-conjugation and the $\dagger$-conjugation, restricted to $\mathbb{C}(\mathbf{j})$, coincide with the usual conjugation there. In order to avoid any confusion with the notation, from now on we will identify the conjugation on $\mathbb{C}(\mathbf{j})$ with the $*$-conjugation. Note also that any element in $\mathbb{C}(\mathbf{j})$ is fixed by the bar-conjugation.

Finally, if $Z = x_1 + \mathbf{ij}\,y_2 \in \mathbb{D}$, that is, $y_1 = x_2 = 0$, then

$$\overline{Z} = x_1 - \mathbf{ij}\,y_2 = Z^\dagger, \qquad Z^* = Z.$$

Thus, the bar-conjugation and the $\dagger$-conjugation restricted to $\mathbb{D}$ coincide with the intrinsic conjugation there. We will use the bar-conjugation to denote the latter. Note that any hyperbolic number is fixed b the $*$-conjugation.

Combining formulas (1.1) and the three conjugations we give now the alternative, although, of course, equivalent way for presenting the conjugates of a bicomplex number Z:

(I') $\overline{Z} = \eta_1 - \eta_2\,\mathbf{i} = \bar{\mathfrak{z}}_1 + \mathbf{j}\,\bar{\mathfrak{z}}_2$
$\quad\;\; = \bar{\xi}_1 - \mathbf{i}\,\bar{\xi}_2 = \bar{\alpha}_1 - \mathbf{k}\,\bar{\alpha}_2 = v_1 - \mathbf{k}\,v_2;$

(II') $Z^\dagger = \eta_1^* + \eta_2^*\,\mathbf{i} = \bar{\mathfrak{z}}_1 - \mathbf{j}\,\bar{\mathfrak{z}}_2$
$\quad\;\; = \bar{\xi}_1 + \mathbf{i}\,\bar{\xi}_2 = \alpha_1 - \mathbf{k}\,\alpha_2 = v_1^* - \mathbf{k}\,v_2^*;$

(III') $Z^* = \eta_1^* - \eta_2^*\,\mathbf{i} = \mathfrak{z}_1 - \mathbf{j}\,\mathfrak{z}_2$
$\quad\;\; = \xi_1 - \mathbf{i}\,\xi_2 = \bar{\alpha}_1 + \mathbf{k}\,\bar{\alpha}_2 = v_1^* + \mathbf{k}\,v_2^*.$

Each conjugation is an additive, involutive, and multiplicative operation on $\mathbb{BC}$:

(I) $\overline{(Z + W)} = \overline{Z} + \overline{W}; (Z + W)^\dagger = Z^\dagger + W^\dagger;$
$\quad (Z + W)^* = Z^* + W^*.$

(II) $\overline{\overline{Z}} = Z; \left(Z^\dagger\right)^\dagger = Z; (Z^*)^* = Z.$

(III) $\overline{(Z \cdot W)} = \overline{Z} \cdot \overline{W}; (Z \cdot W)^\dagger = Z^\dagger \cdot W^\dagger; (Z \cdot W)^* = Z^* \cdot W^*.$

Thus, each conjugation is a ring automorphism of $\mathbb{BC}$.

In the complex case the modulus of a complex number is intimately related with the complex conjugation: multiplying a complex number by its conjugate one gets the square of its modulus. Applying this idea to each of the three conjugations, three possible "moduli" arise in accordance with the formulas for their squares:

- $|Z|_{\mathbf{i}}^2 := Z \cdot Z^\dagger = z_1^2 + z_2^2$

$$= \left(|\eta_1|^2 - |\eta_2|^2\right) + 2\,Re\,(\eta_1\,\eta_2^*)\,\mathbf{i}$$

$$= \left(|\mathfrak{z}_1|^2 + |\mathfrak{z}_2|^2\right) + \mathbf{j}\left(\bar{\mathfrak{z}}_1\,\mathfrak{z}_2 - \overline{\mathfrak{z}_1\,\mathfrak{z}_2}\right)$$

$$= \left(|\mathfrak{r}_1|^2 - |\mathfrak{r}_2|^2 \right) + \mathbf{i} \left(\mathfrak{r}_1 \bar{\mathfrak{r}}_2 + \overline{\mathfrak{r}_1 \bar{\mathfrak{r}}_2} \right)$$

$$= \alpha_1^2 - \alpha_2^2$$

$$= \left(|v_1|^2 - |v_2|^2 \right) - \mathbf{i}\, 2\, Im \left(v_1^* v_2 \right) \in \mathbb{C}(\mathbf{i});$$

- $|Z|_{\mathbf{j}}^2 := Z \cdot \overline{Z} = \left(|z_1|^2 - |z_2|^2 \right) + 2\, Re \left(z_1 \bar{z}_2 \right) \mathbf{j}$

$$= \eta_1^2 + \eta_2^2$$

$$= \left(|\mathfrak{z}_1|^2 - |\mathfrak{z}_2|^2 \right) + \mathbf{j} \left(\mathfrak{z}_1 \bar{\mathfrak{z}}_2 + \overline{\mathfrak{z}_1 \bar{\mathfrak{z}}_2} \right)$$

$$= \left(|\mathfrak{r}_1|^2 + |\mathfrak{r}_2|^2 \right) + \mathbf{i} \left(\bar{\mathfrak{r}}_1 \mathfrak{r}_2 - \overline{\bar{\mathfrak{r}}_1 \mathfrak{r}_2} \right)$$

$$= \left(|\alpha_1|^2 - |\alpha_2|^2 \right) + \mathbf{k} \left(\alpha_2 \bar{\alpha}_1 - \alpha_1 \bar{\alpha}_2 \right)$$

$$= v_1^2 - v_2^2 \in \mathbb{C}(\mathbf{j});$$

- $|Z|_{\mathbf{k}}^2 := Z \cdot Z^* = \left(|z_1|^2 + |z_2|^2 \right) - 2\, Im \left(z_1 \bar{z}_2 \right) \mathbf{k}$

$$= \left(|\eta_1|^2 + |\eta_2|^2 \right) - 2\, Im \left(\eta_1 \eta_2^* \right) \mathbf{k}$$

$$= \mathfrak{z}_1^2 + \mathfrak{z}_2^2$$

$$= \mathfrak{r}_1^2 + \mathfrak{r}_2^2$$

$$= \left(|\alpha_1|^2 + |\alpha_2|^2 \right) + \mathbf{k} \left(\alpha_2 \bar{\alpha}_1 + \alpha_1 \bar{\alpha}_2 \right)$$

$$= \left(|v_1|^2 + |v_2|^2 \right) + \mathbf{k} \left(v_1 v_2^* + v_2 v_1^* \right) \in \mathbb{D},$$

where for a complex number z (in $\mathbb{C}(\mathbf{i})$ or $\mathbb{C}(\mathbf{j})$) we denote by $|z|$ its usual modulus and for a hyperbolic number $\mathfrak{z} = a + b\,\mathbf{k}$ we use the notation $|\mathfrak{z}|^2 = a^2 - b^2$.

Unlike what happens in the complex case, these moduli are not $\mathbb{R}^+$-valued. The first two moduli are complex-valued (in $\mathbb{C}(\mathbf{i})$ and $\mathbb{C}(\mathbf{j})$ respectively), while the last one is hyperbolic-valued. These moduli nevertheless behave as expected with respect to multiplication. Specifically, we have

$$|Z \cdot W|_{\mathbf{i}}^2 = |Z|_{\mathbf{i}}^2 \cdot |W|_{\mathbf{i}}^2;$$

$$|Z \cdot W|_{\mathbf{j}}^2 = |Z|_{\mathbf{j}}^2 \cdot |W|_{\mathbf{j}}^2;$$

$$|Z \cdot W|_{\mathbf{k}}^2 = |Z|_{\mathbf{k}}^2 \cdot |W|_{\mathbf{k}}^2.$$

Remark 1.1 The hyperbolic-valued modulus $|Z|_{\mathbf{k}}$ of a bicomplex number Z satisfies

$$|Z|_{\mathbf{k}}^2 = \left(|z_1|^2 + |z_2|^2\right) + (-2\,Im\,(z_1\,\bar{z}_2))\,\mathbf{k} =: a + b\,\mathbf{k},$$

where a and b satisfy the inequalities

$$a^2 - b^2 \geq 0 \quad \text{and} \quad a \geq 0$$

(this is a consequence of the fact that $|\,Im\,(z_1\,\bar{z}_2)\,| \leq |z_1| \cdot |z_2|$).

This remark justifies the introduction of the set of "positive" hyperbolic numbers:

$$\mathbb{D}^+ := \left\{ a + b\,\mathbf{k} \mid a^2 - b^2 \geq 0,\ a \geq 0 \right\},$$

so that $|Z|_{\mathbf{k}}^2 \in \mathbb{D}^+$. Such a definition of "positiveness" for hyperbolic numbers does not look intuitively clear but later we give another description of $\mathbb{D}^+$ that clarifies the reason for such a name; see, for instance [2]. It turns out that the positive hyperbolic numbers play with respect to all hyperbolic numbers a role similar to that of real nonnegative numbers with respect to all real numbers.

1.3 The Euclidean Norm on $\mathbb{BC}$

Since none of the moduli above is real valued, we can consider also the Euclidean norm on $\mathbb{BC}$ seen as $\mathbb{C}^2(\mathbf{i}) = \{\,(z_1, z_2) \mid z_1 + z_2\,\mathbf{j} \in \mathbb{BC}\,\}$, as $\mathbb{C}^2(\mathbf{j}) = \{\,(\eta_1, \eta_2) \mid \eta_1 + \eta_2\,\mathbf{i} \in \mathbb{BC}\,\}$, or as $\mathbb{R}^4 = \{\,(x_1, y_1, x_2, y_2) \mid x_1 + \,+ \mathbf{i}y_1 + \mathbf{j}x_2 + \mathbf{k}y_2 \in \mathbb{BC}\,\}$. This Euclidean norm is connected to the properties of bicomplex numbers via the $\mathbb{D}^+$-valued modulus as follows:

$$|Z| = \sqrt{x_1^2 + y_1^2 + x_2^2 + y_2^2} = \sqrt{|z_1|^2 + |z_2|^2}$$

$$= \sqrt{|\eta_1|^2 + |\eta_2|^2} = \sqrt{Re\left(|Z|_{\mathbf{k}}^2\right)}\,.$$

It is easy to prove (using the triangle inequality) that, for any Z and U in $\mathbb{BC}$,

$$|Z \cdot U| \leq \sqrt{2}\,|Z| \cdot |U|\,. \tag{1.2}$$

We can actually show that if $U \in \mathbb{BC}$ is arbitrary but Z is either a complex number in $\mathbb{C}(\mathbf{i})$ or $\mathbb{C}(\mathbf{j})$, or a hyperbolic number then we can say something more:

a) if $Z \in \mathbb{C}(\mathbf{i})$ or $\mathbb{C}(\mathbf{j})$ then $|Z \cdot U| = |Z| \cdot |U|$;
b) if $Z \in \mathbb{D}$ then

$$|Z \cdot U|^2 = |Z|^2 \cdot |U|^2 + 4\,x_1\,y_2\,Re(\mathbf{i}\,u_1\,\bar{u}_2)$$

where $Z = x_1 + \mathbf{k} y_2$ and $U = u_1 + \mathbf{j} u_2$.
We will prove these two properties in Sect. 1.4.

1.4 Idempotent Decompositions

Since for any bicomplex number $Z = z_1 + z_2 \mathbf{j}$ it is

$$Z \cdot Z^\dagger = z_1^2 + z_2^2 \in \mathbb{C}(\mathbf{i}),$$

it follows that any bicomplex number Z with $|Z|_\mathbf{i} \neq 0$ is invertible, and its inverse is given by

$$Z^{-1} = \frac{Z^\dagger}{|Z|_\mathbf{i}^2}.$$

If, on the other hand, $Z \neq 0$ but $|Z|_\mathbf{i} = 0$ then Z is a zero divisor. In fact, there are no other zero divisors.

We denote the set of zero divisors by $\mathfrak{S}$, thus

$$\mathfrak{S} := \left\{ Z \mid Z \neq 0, \ z_1^2 + z_2^2 = 0 \right\}.$$

It turns out that there are two very special zero divisors. Set

$$\mathbf{e} := \frac{1}{2} \left(1 + \mathbf{i}\mathbf{j} \right),$$

then its †-conjugate is

$$\mathbf{e}^\dagger = \frac{1}{2} \left(1 - \mathbf{i}\mathbf{j} \right).$$

It is immediate to check that

$$\mathbf{e}^2 = \mathbf{e}; \qquad \left(\mathbf{e}^\dagger \right)^2 = \mathbf{e}^\dagger; \qquad \mathbf{e} + \mathbf{e}^\dagger = 1;$$

$$\mathbf{e}^* = \mathbf{e}, \qquad \left(\mathbf{e}^\dagger \right)^* = \mathbf{e}^\dagger; \qquad \mathbf{e} \cdot \mathbf{e}^\dagger = 0.$$

The last property says that $\mathbf{e}$ and $\mathbf{e}^\dagger$ are zero divisors, and the first three mean that they are mutually complementary idempotent elements.

Thus, the two sets

$$\mathbb{BC}_\mathbf{e} := \mathbf{e} \cdot \mathbb{BC} \quad \text{and} \quad \mathbb{BC}_{\mathbf{e}^\dagger} := \mathbf{e}^\dagger \cdot \mathbb{BC}$$

are (principal) ideals in the ring $\mathbb{BC}$ and they have the properties:

$$\mathbb{BC}_\mathbf{e} \cap \mathbb{BC}_{\mathbf{e}^\dagger} = \{0\}$$

and

$$\mathbb{BC} = \mathbb{BC}_\mathbf{e} + \mathbb{BC}_{\mathbf{e}^\dagger} . \tag{1.3}$$

We shall call (1.3) the idempotent decomposition of $\mathbb{BC}$, and we shall see later that bicomplex modules inherit from their scalars a similar decomposition. Of course, both ideals $\mathbb{BC}_\mathbf{e}$ and $\mathbb{BC}_{\mathbf{e}^\dagger}$ are uniquely determined but their elements admit different representations. In fact, every bicomplex number $Z = (x_1 + \mathbf{i}\, y_1) + (x_2 + \mathbf{i}\, y_2) = z_1 + z_2\,\mathbf{j}$ can be written as

$$Z = z_1 + z_2\,\mathbf{j} = \beta_1\,\mathbf{e} + \beta_2\,\mathbf{e}^\dagger , \tag{1.4}$$

where

$$\beta_1 = z_1 - \mathbf{i}\, z_2 \quad \text{and} \quad \beta_2 = z_1 + \mathbf{i}\, z_2, \tag{1.5}$$

are complex numbers in $\mathbb{C}(\mathbf{i})$. On the other hand, Z can also be written as

$$Z = \eta_1 + \eta_2\,\mathbf{i} = \gamma_1\,\mathbf{e} + \gamma_2\,\mathbf{e}^\dagger , \tag{1.6}$$

where $\eta_1 := x_1 + x_2\,\mathbf{j}$, $\eta_2 := y_1 + y_2\,\mathbf{j}$, $\gamma_1 := \eta_1 - \mathbf{j}\,\eta_2$, $\gamma_2 = \eta_1 + \mathbf{j}\,\eta_2$ are complex numbers in $\mathbb{C}(\mathbf{j})$. Each of the formulas (1.4) and (1.6) can be equally called the idempotent representation of a bicomplex number. More specifically, (1.4) is the idempotent representation for $\mathbb{BC}$ seen as $\mathbb{C}^2(\mathbf{i}) := \mathbb{C}(\mathbf{i}) \times \mathbb{C}(\mathbf{i})$ and (1.6) is the idempotent representation for $\mathbb{BC}$ seen as $\mathbb{C}^2(\mathbf{j}) := \mathbb{C}(\mathbf{j}) \times \mathbb{C}(\mathbf{j})$. Usually, (see for instance [1–3]) only the representation (1.4) is considered, but we see no reason to restrict ourselves just to this case: the consequences are similar but different.

It is worth pointing out that

$$\beta_1\,\mathbf{e} = \gamma_1\,\mathbf{e} \quad \text{and} \quad \beta_2\,\mathbf{e}^\dagger = \gamma_2\,\mathbf{e}^\dagger$$

although β_1 and β_2 are in $\mathbb{C}(\mathbf{i})$, while γ_1 and γ_2 are in $\mathbb{C}(\mathbf{j})$. More specifically, given $\beta_1 = Re\,\beta_1 + \mathbf{i}\,Im\,\beta_1$, the equality

$$\beta_1\,\mathbf{e} = \gamma_1\,\mathbf{e}$$

is true if and only if

$$\gamma_1 = Re\,\beta_1 - \mathbf{j}\,Im\,\beta_1.$$

Similarly if $\beta_2 = Re\,\beta_2 + \mathbf{i}\,Im\,\beta_2$, the equality

$$\beta_2\,\mathbf{e}^\dagger = \gamma_2\,\mathbf{e}^\dagger$$

is true if and only if

$$\gamma_2 = Re\,\beta_2 + \mathbf{j}\,Im\,\beta_2.$$

Altogether, decomposition (1.3) can be written in any of the two equivalent forms:

$$\mathbb{BC} = \mathbb{C}(\mathbf{i}) \cdot \mathbf{e} + \mathbb{C}(\mathbf{i}) \cdot \mathbf{e}^\dagger\,;$$

$$\mathbb{BC} = \mathbb{C}(\mathbf{j}) \cdot \mathbf{e} + \mathbb{C}(\mathbf{j}) \cdot \mathbf{e}^\dagger\,.$$

If $\mathbb{BC}$ is seen as a $\mathbb{C}(\mathbf{i})$-linear (respectively, a $\mathbb{C}(\mathbf{j})$-linear) space then the first (respectively, the second) decomposition becomes a direct sum.

It is usually stated that the idempotent representation is unique which seems to contradict our two formulas. But the fact is that each of (1.4) and (1.6) is unique in the following sense. Assuming that a bicomplex number Z has two idempotent representations with coefficients in $\mathbb{C}(\mathbf{i})$: $Z = \beta_1\,\mathbf{e} + \beta_2\,\mathbf{e}^\dagger = \eta_1\,\mathbf{e} + \eta_2\,\mathbf{e}^\dagger$ with $\beta_1, \beta_2, \eta_1, \eta_2 \in \mathbb{C}(\mathbf{i})$ then it is easy to show that $\beta_1 = \eta_1$ and $\beta_2 = \eta_2$; similarly $Z = \gamma_1\,\mathbf{e} + \gamma_2\,\mathbf{e}^\dagger = \xi_1\,\mathbf{e} + \xi_2\,\mathbf{e}^\dagger$ with $\gamma_1, \gamma_2, \xi_1, \xi_2 \in \mathbb{C}(\mathbf{j})$ implies that $\gamma_1 = \xi_1$ and $\gamma_2 = \xi_2$.

Note that formula (1.4) gives a relation between the two bases $\{1, \mathbf{j}\}$ and $\{\mathbf{e}, \mathbf{e}^\dagger\}$ in the $\mathbb{C}(\mathbf{i})$-linear space $\mathbb{BC} = \mathbb{C}^2(\mathbf{i})$. One has the following transition formula using the matrix of change of variables:

$$\begin{pmatrix} z_1 \\ z_2 \end{pmatrix} = \begin{pmatrix} \dfrac{1}{2} & \dfrac{1}{2} \\[2mm] -\dfrac{1}{2\mathbf{i}} & \dfrac{1}{2\mathbf{i}} \end{pmatrix} \cdot \begin{pmatrix} \beta_1 \\ \beta_2 \end{pmatrix}. \tag{1.7}$$

Note that the new basis $\{\mathbf{e}, \mathbf{e}^\dagger\}$ is orthogonal with respect to the Euclidean inner product in $\mathbb{C}^2(\mathbf{i})$ given by

$$\langle (z_1, z_2)\,,\,(u_1, u_2) \rangle_{\mathbb{C}^2(\mathbf{i})} := z_1\overline{u}_1 + z_2\overline{u}_2$$

for all $(z_1, z_2),\,(u_1, u_2) \in \mathbb{C}^2(\mathbf{i})$. Indeed, since $\mathbf{e} = \left(\dfrac{1}{2}, \dfrac{\mathbf{i}}{2}\right)$ and $\mathbf{e}^\dagger = \left(\dfrac{1}{2}, -\dfrac{\mathbf{i}}{2}\right)$ as elements in $\mathbb{C}^2(\mathbf{i})$, we have:

$$\left\langle \mathbf{e}, \mathbf{e}^\dagger \right\rangle_{\mathbb{C}^2(\mathbf{i})} = 0 \quad\text{and}\quad \langle \mathbf{e}, \mathbf{e} \rangle_{\mathbb{C}^2(\mathbf{i})} = \left\langle \mathbf{e}^\dagger, \mathbf{e}^\dagger \right\rangle_{\mathbb{C}^2(\mathbf{i})} = \frac{1}{2},$$

and so $\{\mathbf{e}, \mathbf{e}^\dagger\}$ is an orthogonal but not orthonormal basis for $\mathbb{C}^2(\mathbf{i})$. As a consequence one gets:

$$|Z| = \frac{1}{\sqrt{2}}\sqrt{|\beta_1|^2 + |\beta_2|^2}\,. \tag{1.8}$$

Formula (1.6) can be interpreted analogously. Now we have $\mathbb{BC} = \mathbb{C}^2(\mathbf{j})$ and two bases in it as $\mathbb{C}(\mathbf{j})$-linear space are $\{1, \mathbf{i}\}$ and again(!) $\{\mathbf{e}, \mathbf{e}^\dagger\}$.

We emphasize here how the versatile character of the idempotents $\mathbf{e}$ and $\mathbf{e}^\dagger$ is manifested: for representation (1.4) the bicomplex number $\mathbf{e}$ should be seen as $\mathbb{BC} \ni \mathbf{e} = z_1 + z_2 \mathbf{j} = \frac{1}{2} + \left(\frac{1}{2}\mathbf{i}\right)\mathbf{j}$, that is, in this case $z_1 = \frac{1}{2}, z_2 = \frac{1}{2}\mathbf{i}$; for representation (1.6) the idempotent element $\mathbf{e}$ should be seen as $\mathbb{BC} \ni \mathbf{e} = \eta_1 + \eta_2 \mathbf{i} = \frac{1}{2} + \left(\frac{1}{2}\mathbf{j}\right)\mathbf{i}$, with $\eta_1 = \frac{1}{2}, \eta_2 = \frac{1}{2}\mathbf{j}$.

Note that $\mathbf{e}$ and $\mathbf{e}^\dagger$ form a basis in both $\mathbb{C}^2(\mathbf{i})$ and $\mathbb{C}^2(\mathbf{j})$. The basis $\{\mathbf{e}, \mathbf{e}^\dagger\}$ remains orthogonal in $\mathbb{C}^2(\mathbf{j})$ with transition matrix

$$\begin{pmatrix} \dfrac{1}{2} & \dfrac{1}{2} \\[2mm] -\dfrac{1}{2\mathbf{j}} & \dfrac{1}{2\mathbf{j}} \end{pmatrix}$$

and if $Z = \gamma_1 \mathbf{e} + \gamma_2 \mathbf{e}^\dagger$, $\gamma_1, \gamma_2 \in \mathbb{C}(\mathbf{j})$, then

$$|Z| = \frac{1}{\sqrt{2}}\sqrt{|\gamma_1|^2 + |\gamma_2|^2}\,. \tag{1.9}$$

A major advantage of both idempotent representations of bicomplex numbers is that they allow to perform the operations of addition, multiplication, inverse, square roots, etc., component-wise. For more details see, [2].

Let us come back to properties $a)$ and $b)$ in Sect. 1.3. We prove first $a)$. Indeed, take $Z = z_1 \in \mathbb{C}(\mathbf{i})$ and $U = u_1 + u_2 \mathbf{j} = (u_1 - \mathbf{i} u_2)\mathbf{e} + (u_1 + \mathbf{i} u_2)\mathbf{e}^\dagger$, then

$$|Z \cdot U|^2 = |z_1 (u_1 + u_2 \mathbf{j})|^2 = |(z_1 u_1) + (z_1 u_2)\mathbf{j}|^2$$
$$= |z_1 u_1|^2 + |z_1 u_2|^2 = |z_1|^2 \cdot |U|^2,$$

and the first part of property $a)$ is proved. As for the second part, take $Z = x_1 + x_2 \mathbf{j} = (x_1 - \mathbf{i} x_2)\mathbf{e} + (x_1 + \mathbf{i} x_2)\mathbf{e}^\dagger \in \mathbb{C}(\mathbf{j})$, then

$$|Z \cdot U|^2 = \left| \left((x_1 - \mathbf{i} x_2)\mathbf{e} + (x_1 + \mathbf{i} x_2)\mathbf{e}^\dagger\right) \cdot \right.$$
$$\left. \cdot \left((u_1 - \mathbf{i} u_2)\mathbf{e} + (u_1 + \mathbf{i} u_2)\mathbf{e}^\dagger\right) \right|^2$$
$$= |(x_1 - \mathbf{i} x_2)(u_1 - \mathbf{i} u_2)\mathbf{e} + (x_1 + \mathbf{i} x_2)(u_1 + \mathbf{i} u_2)\mathbf{e}^\dagger|^2$$
$$= \frac{1}{2}\left(|x_1 - \mathbf{i}x_2|^2 \cdot |u_1 - \mathbf{i}u_2|^2 + |x_1 + \mathbf{i}x_2|^2 \cdot |u_1 + \mathbf{i}u_2|^2\right)$$
$$= |Z|^2 \cdot |U|^2,$$

thus property a) is proved. Finally, in order to deal with property b), take $Z = x_1 + y_2 \mathbf{i}\mathbf{j} = (x_1 + y_2)\mathbf{e} + (x_1 - y_2)\mathbf{e}^\dagger \in \mathbb{D}$, then

$$|Z \cdot U|^2 = |(x_1 + y_2 \mathbf{i}\mathbf{j}) \cdot (u_1 + u_2 \mathbf{j})|^2$$

$$= \left| (x_1 + y_2) \cdot (u_1 - \mathbf{i}u_2)\mathbf{e} + (x_1 - y_2) \cdot (u_1 + \mathbf{i}u_2)\mathbf{e}^\dagger \right|^2$$

$$= \frac{1}{2} \left((x_1 + y_2)^2 \cdot |u_1 - \mathbf{i}u_2|^2 + (x_1 - y_2)^2 \cdot |u_1 + \mathbf{i}u_2|^2 \right)$$

$$= |Z|^2 \cdot |U|^2 + 4 x_1 y_2 \, Re(\mathbf{i} u_1 \bar{u}_2) \,.$$

In the following sections we will need the mappings:

$$\pi_{1,\mathbf{i}}, \; \pi_{2,\mathbf{i}} : \mathbb{BC} \to \mathbb{C}(\mathbf{i})$$

given by

$$\pi_{\ell,\mathbf{i}}(Z) = \pi_{\ell,\mathbf{i}}(\beta_1 \mathbf{e} + \beta_2 \mathbf{e}^\dagger) := \beta_\ell \in \mathbb{C}(\mathbf{i}) \,.$$

These maps are nothing but the projections onto the "coordinate axes" in $\mathbb{C}^2(\mathbf{i})$ with the basis $\{\mathbf{e}, \mathbf{e}^\dagger\}$. Completely analogously one has the mappings

$$\pi_{1,\mathbf{j}}, \; \pi_{2,\mathbf{j}} : \mathbb{BC} \to \mathbb{C}(\mathbf{j})$$

given by

$$\pi_{\ell,\mathbf{j}}(Z) = \pi_{\ell,\mathbf{j}}(\gamma_1 \mathbf{e} + \gamma_2 \mathbf{e}^\dagger) := \gamma_\ell \in \mathbb{C}(\mathbf{j}) \,,$$

which are now the projections onto the coordinate axes in $\mathbb{C}^2(\mathbf{j})$ with the same basis $\{\mathbf{e}, \mathbf{e}^\dagger\}$.

Finally, using the cartesian representations for any $Z \in \mathbb{BC}$:

$$Z = (x_1 + \mathbf{i}x_2) + (y_1 + \mathbf{i}y_2)\mathbf{j} = z_1 + z_2\mathbf{j}$$

$$= (x_1 + \mathbf{j}y_1) + (x_2 + \mathbf{j}y_2)\mathbf{i} = \eta_1 + \eta_2\mathbf{i} \,,$$

we define two more projections

$$\Pi_{1,\mathbf{i}}, \; \Pi_{2,\mathbf{i}} : \mathbb{BC} \to \mathbb{C}(\mathbf{i}) \quad \text{and} \quad \Pi_{1,\mathbf{j}}, \; \Pi_{2,\mathbf{j}} : \mathbb{BC} \to \mathbb{C}(\mathbf{j})$$

given by

$$\Pi_{\ell,\mathbf{i}}(Z) = \Pi_{\ell,\mathbf{i}}(z_1 + z_2\mathbf{j}) := z_\ell \in \mathbb{C}(\mathbf{i}) \,,$$

and

$$\Pi_{\ell,\mathbf{j}}(Z) = \Pi_{\ell,\mathbf{j}}(\eta_1 + \eta_2 \mathbf{i}) := \eta_\ell \in \mathbb{C}(\mathbf{j}).$$

The properties of the operations with bicomplex numbers immediately imply the following equalities for any Z, $W \in \mathbb{BC}$:

(a) $\pi_{\ell,\mathbf{i}}(ZW) = \pi_{\ell,\mathbf{i}}(Z)\, \pi_{\ell,\mathbf{i}}(W)$;

(b) $\pi_{\ell,\mathbf{j}}(ZW) = \pi_{\ell,\mathbf{j}}(Z)\, \pi_{\ell,\mathbf{j}}(W)$;

(c) $\forall\, \lambda \in \mathbb{C}(\mathbf{i}),\quad \pi_{\ell,\mathbf{i}}(\lambda) = \lambda; \quad \Pi_{1,\mathbf{i}}(\lambda) = \lambda, \quad \Pi_{2,\mathbf{i}}(\lambda) = 0$;

(d) $\forall\, \mu \in \mathbb{C}(\mathbf{j}),\quad \pi_{\ell,\mathbf{j}}(\mu) = \mu; \quad \Pi_{1,\mathbf{j}}(\mu) = \mu, \quad \Pi_{2,\mathbf{j}}(\mu) = 0$;

(e) $\pi_{1,\mathbf{i}} = \Pi_{1,\mathbf{i}} - \mathbf{i}\,\Pi_{2,\mathbf{i}}; \quad \pi_{2,\mathbf{i}} = \Pi_{1,\mathbf{i}} + \mathbf{i}\,\Pi_{2,\mathbf{i}}$;

(f) $\pi_{1,\mathbf{j}} = \Pi_{1,\mathbf{j}} - \mathbf{j}\,\Pi_{2,\mathbf{j}}; \quad \pi_{2,\mathbf{j}} = \Pi_{1,\mathbf{j}} + \mathbf{j}\,\Pi_{2,\mathbf{j}}$;

(g) $\Pi_{1,\mathbf{i}} = \dfrac{1}{2}\left(\pi_{1,\mathbf{i}} + \pi_{2,\mathbf{i}}\right); \quad \Pi_{2,\mathbf{i}} = \dfrac{\mathbf{i}}{2}\left(\pi_{1,\mathbf{i}} - \pi_{2,\mathbf{i}}\right)$;

(h) $\Pi_{1,\mathbf{j}} = \dfrac{1}{2}\left(\pi_{1,\mathbf{j}} + \pi_{2,\mathbf{j}}\right); \quad \Pi_{2,\mathbf{j}} = \dfrac{\mathbf{j}}{2}\left(\pi_{1,\mathbf{j}} - \pi_{2,\mathbf{j}}\right)$.

We leave the verification of these properties to the reader.

1.5 A Partial Order on $\mathbb{D}$ and a Hyperbolic-Valued Norm

Let us describe some properties that hyperbolic numbers inherit from bicomplex ones. First of all we note that both $\mathbf{e}$ and $\mathbf{e}^\dagger$ are hyperbolic numbers. More importantly, the idempotent representation of any hyperbolic number $\alpha = a + b\,\mathbf{k}$ is

$$\alpha = \nu\, \mathbf{e} + \mu\, \mathbf{e}^\dagger, \quad \nu, \mu \in \mathbb{R},$$

with $\nu = b + a$, $\mu = b - a$. It is also necessary to see that

$$\mathbb{D}^+ = \left\{ \nu\, \mathbf{e} + \mu\, \mathbf{e}^\dagger \mid \nu, \mu \geq 0 \right\}.$$

Thus positive hyperbolic numbers are those whose both idempotent components are nonnegative, which somehow explains the origin of the name.

In Fig. 1.1 the points (x, y) correspond to the hyperbolic numbers $Z = x + \mathbf{k}y$. One sees that, geometrically, the hyperbolic positive numbers are situated in the quarter plane denoted by $\mathbb{D}^+$. The quarter plane symmetric to it with respect to the origin corresponds to the "negative" hyperbolic numbers, i.e., to those that have both idempotent components negative. The rest of the points correspond to those hyperbolic numbers that cannot be called either positive or negative.

Let us now define in $\mathbb{D}$ the following binary relation: given $\alpha_1, \alpha_2 \in \mathbb{D}$, we write $\alpha_1 \preccurlyeq \alpha_2$ whenever $\alpha_2 - \alpha_1 \in \mathbb{D}^+$. It is obvious that this relation is reflexive, transitive, and antisymmetric, and that therefore it defines a partial order on $\mathbb{D}$. With abuse of notation, we will say that α_2 is $\mathbb{D}$-larger ($\mathbb{D}$-less respectively) than α_1 if $\alpha_1 \preccurlyeq \alpha_2$ ($\alpha_2 \preccurlyeq \alpha_1$ respectively).

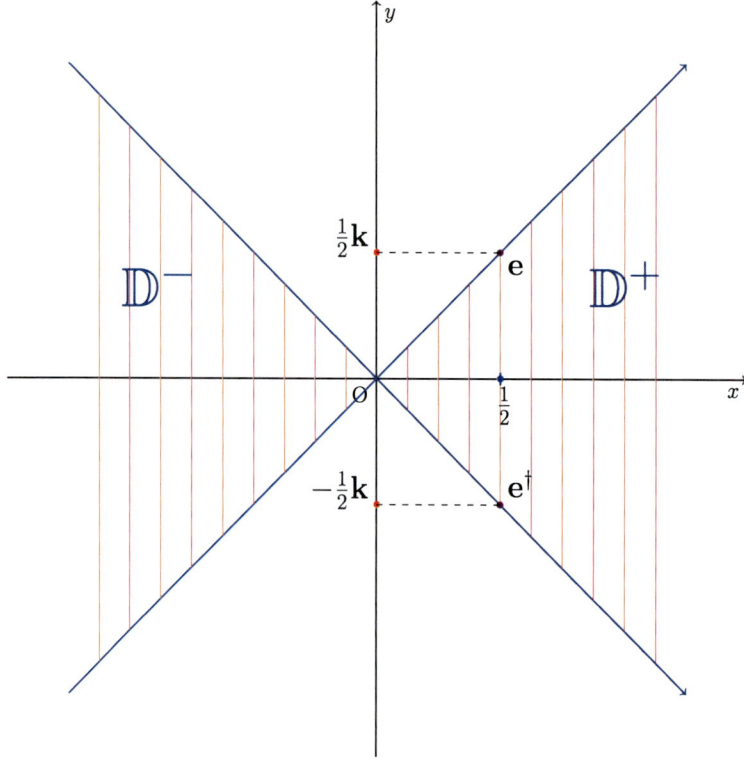

Fig. 1.1 Positive and negative hyperbolic numbers

Observe that given r, $s \in \mathbb{R}$, then $r \leq s$ if and only if $r \prec s$, that is, the partial order $\prec$ is an extension of the total order $\leq$ in $\mathbb{R}$. In fact, $\prec$ defines a total order on any line through the origin in $\mathbb{D}$.

In Fig. 1.2, $Z_0 = x_0 + \mathbf{k}y_0$ is an arbitrary hyperbolic number, and one can see that the entire plane is divided into: the quarter plane of hyperbolic numbers which are $\mathbb{D}$-greater than Z_0 ($Z \succ Z_0$); the quarter plane of hyperbolic numbers which are $\mathbb{D}$-less than Z_0 ($Z \prec Z_0$); and the two quarter planes where the hyperbolic numbers are not $\mathbb{D}$-comparable with Z_0 (neither $Z \succ Z_0$ nor $Z \prec Z_0$ holds).

Remark 1.5.1 While the notion of positive hyperbolic number may not appear immediately intuitive, one can see its relation to a two-dimensional space time with the Minkowksi metric. Specifically, consider a one-dimensional time, say along the $x-$axis, and a one-dimensional space along the $y-$axis. Embed both axes in the hyperbolic space $\mathbb{D}$. Finally, assume the speed of light to be normalized to one. Now, the set of zero-divisors with $x > 0$ represents the future of a ray of light being sent from the origin toward either direction. Analogously, the set of zero-divisors with $x < 0$ represents the past of the same ray of light.

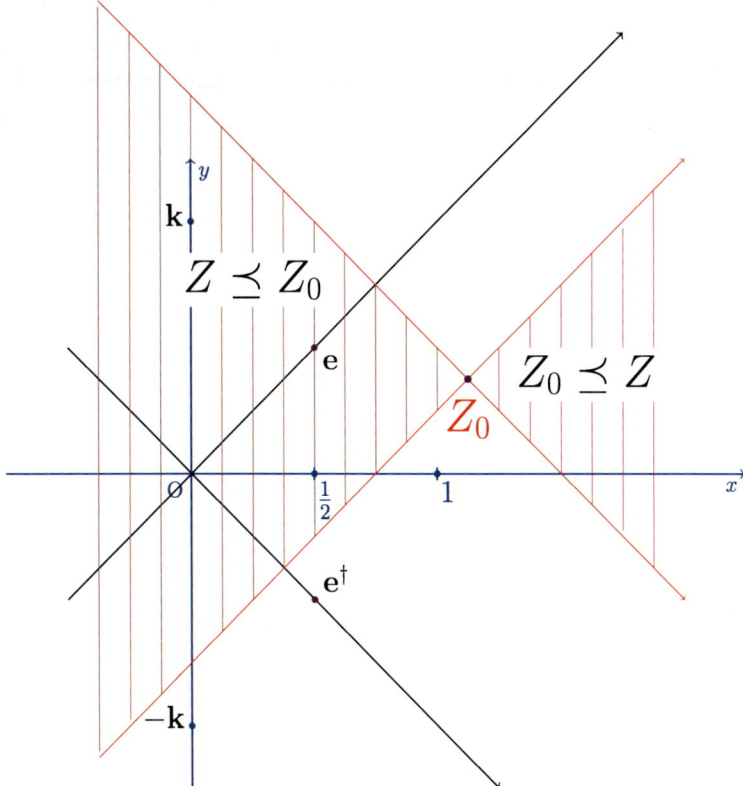

Fig. 1.2 Hyperbolic numbers comparable with a given hyperbolic number

Thus, positive hyperbolic numbers are nothing but the future cone, i.e., they describe the events that are in the future of the origin, while the negative hyperbolic numbers are the past cone, i.e., they are the events in the future of the origin. Within this interpretation, we see how the set of hyperbolic numbers larger than a given hyperbolic number Z represent the events in the future of the event represented by Z.

Given a subset $\mathcal{A}$ in $\mathbb{D}$, we can now define a notion of $\mathbb{D}$-upper and $\mathbb{D}$-lower bound, as well as the notions of a set being $\mathbb{D}$-bounded from above, from below, and finally of a $\mathbb{D}$-bounded set. If $\mathcal{A} \subset \mathbb{D}$ is a set $\mathbb{D}$-bounded from above, we define the notion of its supremum, $sup_{\mathbb{D}}\mathcal{A}$ as usual to be the least upper bound for $\mathcal{A}$. However, it is immediate to note that one can find a more convenient expression for it, as indicated in the following remark:

Remark 1.5.2 Given a set $\mathcal{A} \subset \mathbb{D}$, $\mathbb{D}$-bounded from above, consider the sets $\mathcal{A}_1 := \{\, a_1 \mid a_1\mathbf{e} + a_2\mathbf{e}^\dagger \in \mathcal{A} \,\}$ and $\mathcal{A}_2 := \{\, a_2 \mid a_1\mathbf{e} + a_2\mathbf{e}^\dagger \in \mathcal{A} \,\}$. Then the $sup_{\mathbb{D}} \mathcal{A}$ is given by

$$\sup{}_{\mathbb{D}}\mathcal{A} := \sup \mathcal{A}_1 \cdot \mathbf{e} + \sup \mathcal{A}_2 \cdot \mathbf{e}^\dagger.$$

We have previously defined the hyperbolic modulus $|Z|_{\mathbf{k}}^2$ of any $Z \in \mathbb{BC}$ by the formula $|Z|_{\mathbf{k}}^2 := Z \cdot Z^*$. Thus, writing $Z = \beta_1 \mathbf{e} + \beta_2 \mathbf{e}^\dagger$, one has that $|Z|_{\mathbf{k}}^2 = |\beta_1|^2 \mathbf{e} + |\beta_2|^2 \mathbf{e}^\dagger$, and taking the positive square roots of each coefficient we have that

$$|Z|_{\mathbf{k}} := |\beta_1| \mathbf{e} + |\beta_2| \mathbf{e}^\dagger.$$

That is, we have a map

$$|\cdot|_{\mathbf{k}} : \mathbb{BC} \longrightarrow \mathbb{D}^+$$

with the following properties:

(I) $|Z|_{\mathbf{k}} = 0$ if and only if $Z = 0$;
(II) $|Z \cdot W|_{\mathbf{k}} = |Z|_{\mathbf{k}} \cdot |W|_{\mathbf{k}}$ for any $Z, W \in \mathbb{BC}$;
(III) $|Z + W|_{\mathbf{k}} \leqslant |Z|_{\mathbf{k}} + |W|_{\mathbf{k}}$.

The first two of them are clear. Let us prove (III).

$$|Z + W|_{\mathbf{k}} = |(\beta_1 + \nu_1) \cdot \mathbf{e} + (\beta_2 + \nu_2) \cdot \mathbf{e}^\dagger|_{\mathbf{k}}$$

$$= |\beta_1 + \nu_1| \cdot \mathbf{e} + |\beta_2 + \nu_2| \cdot \mathbf{e}^\dagger$$

$$\leqslant (|\beta_1| + |\nu_1|) \cdot \mathbf{e} + (|\beta_2| + |\nu_2|) \cdot \mathbf{e}^\dagger$$

$$= |Z|_{\mathbf{k}} + |W|_{\mathbf{k}}.$$

It is on the basis of these properties that we will say that $|\cdot|_{\mathbf{k}}$ is the hyperbolic-valued ($\mathbb{D}$-valued) norm on the $\mathbb{BC}$-module $\mathbb{BC}$.

It is instructive to compare (II) with (1.2) where the norm of the product and the product of the norms are related with an inequality. We believe that one could say that the hyperbolic norm of bicomplex numbers is better compatible with the algebraic structure of the latter although, of course, one has to allow hyperbolic values.

Remark 1.5.3 (1) Since for any $Z \in \mathbb{BC}$ it holds that

$$|Z|_{\mathbf{k}} \leqslant \sqrt{2} \cdot |Z| \tag{1.10}$$

with $|Z|$ the Euclidean norm of Z, one has:

$$|Z \cdot W|_{\mathbf{k}} \leqslant \sqrt{2} \cdot |Z| \cdot |W|_{\mathbf{k}}.$$

In contrast with property (II), the above inequality involves both the Euclidean and the hyperbolic norms.

(2) Take γ and ν in $\mathbb{D}^+$, then clearly

$$\gamma \prec \nu \quad \Longrightarrow \quad |\gamma| \le |\nu|. \tag{1.11}$$

(3) Note that the definition of hyperbolic norm for a bicomplex number Z does not depend on the choice of its idempotent representation. We have used, for $Z \in \mathbb{BC}$, the idempotent representation $Z = \beta_1 \mathbf{e} + \beta_2 \mathbf{e}^\dagger$, with $\beta_1, \beta_2 \in \mathbb{C}(\mathbf{i})$. If we would have departed from the idempotent representation $Z = \gamma_1 \mathbf{e} + \gamma_2 \mathbf{e}^\dagger$, with $\gamma_1, \gamma_2 \in \mathbb{C}(\mathbf{j})$ then we would have arrived at the same definition of the hyperbolic norm since $|\beta_1| = |\gamma_1|$ and $|\beta_2| = |\gamma_2|$.

The comparison of the Euclidean norm $|Z|$ and the $\mathbb{D}$-valued norm $|Z|_\mathbf{k}$ of a bicomplex number gives

$$||Z|_\mathbf{k}| = \frac{1}{2} \sqrt{|\beta_1|^2 + |\beta_2|^2} = |Z| \tag{1.12}$$

where the left-hand side is the Euclidean norm of a hyperbolic number.

Now that the norms have been introduced in $\mathbb{BC}$, one is able to talk about convergence. It is clear what the convergence means if it is related with the Euclidean norm. But we are not aware of any work that treats the convergence with respect to a norm with hyperbolic values.

Definition 1.5.4 A sequence of bicomplex numbers $\{Z_n\}_{n \in \mathbb{N}}$ converges to the bicomplex number Z_0 with respect to the hyperbolic-valued norm $| \cdot |_\mathbf{k}$ if for all $\varepsilon > 0$ there exists $N \in \mathbb{N}$ such that for all $n \ge N$ there holds:

$$|Z_n - Z_0|_\mathbf{k} \prec \varepsilon.$$

From inequality (1.10) and equality (1.12), it follows that a sequence $\{Z_n\}_{n \in \mathbb{N}}$ converges to the bicomplex number Z_0 with respect to the hyperbolic-valued norm if and only if it converges to Z_0 with respect to the Euclidean norm, and so even though the two norms cannot be compared as they take values in different rings, one still obtains the same notion of convergence.

It is possible to give a precise geometrical description of the set of bicomplex numbers with a fixed hyperbolic norm, that is, we want to introduce the "sphere of hyperbolic radius γ_0 inside $\mathbb{BC}$". In other words, given $\gamma_0 = a_0 \mathbf{e} + b_0 \mathbf{e}^\dagger$ (see Fig. 1.3) we are looking for the set

$$\mathbb{S}_{\gamma_0} := \{Z \in \mathbb{BC} \mid |Z|_\mathbf{k} = \gamma_0 \}.$$

First note that, if one of a_0 or b_0 is zero, let us say $\gamma_0 = a_0 \cdot \mathbf{e}$, then

$$\mathbb{S}_{\gamma_0} = \{Z = \beta_1 \cdot \mathbf{e} \mid |\beta_1| = a_0 \},$$

and this set is a circumference in the real two–dimensional plane $\mathbb{BC}_\mathbf{e}$ with center at the origin and radius $\dfrac{a_0}{\sqrt{2}} = |a_0 \cdot \mathbf{e}|$. Similarly, if $\gamma_0 = b_0 \cdot \mathbf{e}^\dagger$, the set $\mathbb{S}_{\gamma_0}$ is a circumference in $\mathbb{BC}_{\mathbf{e}^\dagger}$ with center at the origin and radius $\dfrac{b_0}{\sqrt{2}} = |b_0 \cdot \mathbf{e}^\dagger|$.

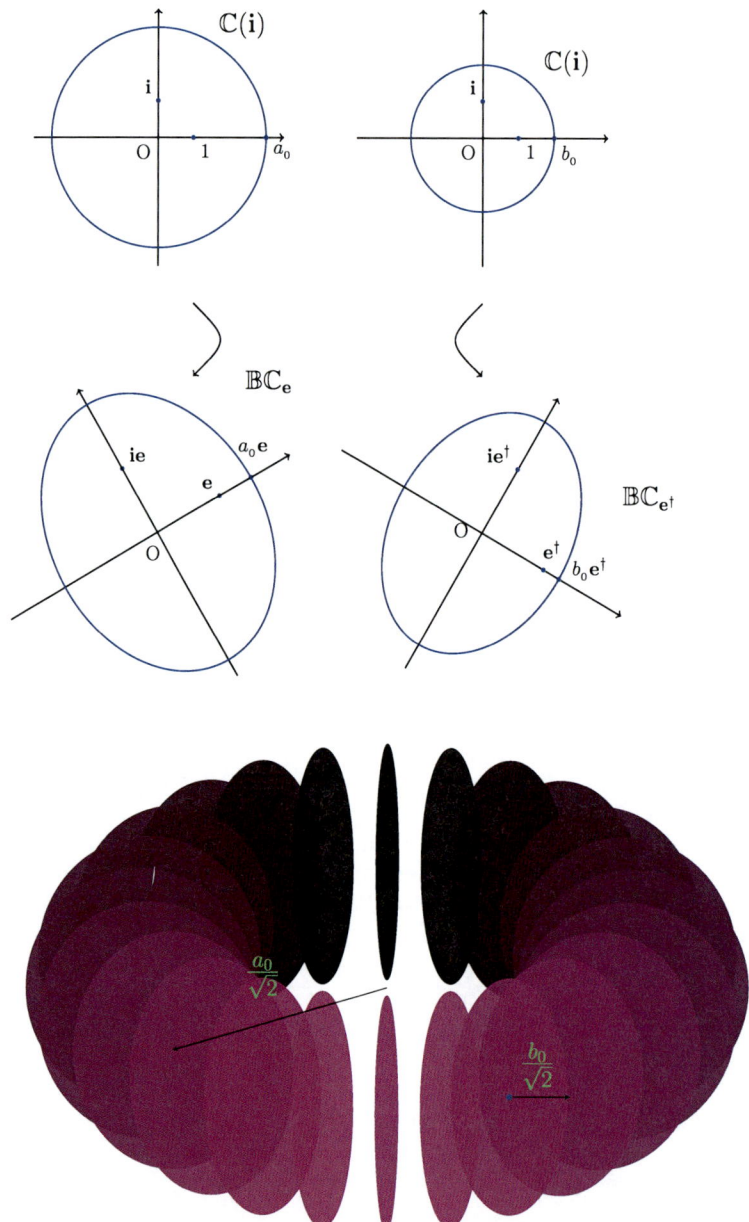

Fig. 1.3 The bicomplex sphere of hyperbolic radius

If $a_0 \neq 0$ and $b_0 \neq 0$ then the intersection of the hyperbolic plane $\mathbb{D}$ and the sphere $\mathbb{S}_{\gamma_0}$ consists of exactly four hyperbolic numbers: $\pm a_0 \mathbf{e} \pm b_0 \mathbf{e}^{\dagger}$, where the plane $\mathbb{D}$ touches the sphere tangentially. What is more, in this case the sphere $\mathbb{S}_{\gamma_0}$ is the surface of a three-dimensional torus, which is obtained by taking the cartesian product of the two circumferences, one in the plane $\mathbb{BC_e}$, and the other in the plane $\mathbb{BC}_{\mathbf{e}^{\dagger}}$, both centered at zero and with radii $\dfrac{a_0}{\sqrt{2}}$ and $\dfrac{b_0}{\sqrt{2}}$ respectively:

$$\mathbb{S}_{\gamma_0} = \left\{ \beta_1 \cdot \mathbf{e} + \beta_2 \cdot \mathbf{e}^{\dagger} \ : |\beta_1| = a_0, \ , \ |\beta_2| = b_0 \right\} .$$

See Fig. 1.3.

References

1. D. Rochon, S, Tremblay, Bicomplex quantum mechanics II: the Hilbert space. Adv. Appl. Cliffod Algebras **16**, 135–157 (2006)
2. D. Rochon, M. Shapiro, On algebraic properties of bicomplex and hyperbolic numbers. Anal. Univ. Oradea. fasc. math. **11**, 71–110 (2004)
3. G.B. Price, *An Introduction to Multicomplex Spaces and Functions.* (Marcel Dekker, New York, 1991)

Chapter 2
Bicomplex Functions and Matrices

Abstract This chapter gives a review of known and new properties of bicomplex holomorphic functions as well as bicomplex analogues of some properties of matrices with real and complex entries.

Keywords Bicomplex holomorphic functions · Complex valued holomorphic functions of two complex variables · Derivative of a bicomplex function · Matrices with bicomplex entries

2.1 Bicomplex Holomorphic Functions

The notion of bicomplex holomorphic functions was introduced a long time ago and we refer the reader to the introduction of the book [1]. The book by itself can serve as a first reading on this subject. For the latest developments see [2] and [3]. In this section we present a summary of the basic facts of the theory that will be used in the last chapter.

The *derivative* $F'(Z_0)$ of the function $F : \Omega \subset \mathbb{BC} \to \mathbb{BC}$ at a point $Z_0 \in \Omega$ is defined to be the limit, if it exists,

$$F'(Z_0) := \lim_{\mathfrak{S}_0 \not\ni H \to 0} \frac{F(Z_0 + H) - F(Z_0)}{H}, \tag{2.1}$$

such that $H = h_1 + \mathbf{j}h_2$ is an invertible bicomplex number. In this case, the function F is called *derivable* at Z_0.

In analogy with holomorphic functions in one complex variable the Cauchy–Riemann type conditions arise where now the complex, not real, partial derivatives participate. More explicitly, if we consider a bicomplex function $F = f_1 + \mathbf{j}f_2$ derivable at Z_0, then we have that the complex partial derivatives

$$F'_{z_1}(Z_0) = \lim_{h_1 \to 0} \frac{F(Z_0 + h_1) - F(Z_0)}{h_1},$$

D. Alpay et al., *Basics of Functional Analysis with Bicomplex Scalars, and Bicomplex Schur Analysis*, SpringerBriefs in Mathematics, DOI: 10.1007/978-3-319-05110-9_2, © The Author(s) 2014

$$F'_{z_2}(Z_0) = \lim_{h_2 \to 0} \frac{F(Z_0 + \mathbf{j}h_2) - F(Z_0)}{h_2},$$

exist and verify the identity:

$$F'(Z_0) = F'_{z_1}(Z_0) = -\mathbf{j}F'_{z_2}(Z_0), \tag{2.2}$$

which is equivalent to the *complex Cauchy-Riemann system* for F (at Z_0):

$$(f_1)'_{z_1}(Z_0) = (f_2)'_{z_2}(Z_0), \qquad (f_1)'_{z_2}(Z_0) = -(f_2)'_{z_1}(Z_0). \tag{2.3}$$

Of course, if F has bicomplex derivative at each point of Ω, we will say that F is a *bicomplex holomorphic*, or $\mathbb{BC}$-holomorphic, function.

For a $\mathbb{BC}$-holomorphic function F, formulas (2.3) imply that

$$\frac{\partial F}{\partial \bar{z}_1}(Z_0) = \frac{\partial F}{\partial \bar{z}_2}(Z_0) = 0, \tag{2.4}$$

i.e.,

$$\frac{\partial f_1}{\partial \bar{z}_1}(Z_0) = \frac{\partial f_1}{\partial \bar{z}_2}(Z_0) = \frac{\partial f_2}{\partial \bar{z}_1}(Z_0) = \frac{\partial f_2}{\partial \bar{z}_2}(Z_0) = 0, \tag{2.5}$$

where the symbols $\dfrac{\partial}{\partial \bar{z}_1}$ and $\dfrac{\partial}{\partial \bar{z}_2}$ are the commonly used formal operations on functions of z_1 and z_2.

This means in particular that F is holomorphic with respect to z_1 for any z_2 fixed and F is holomorphic with respect to z_2 for any z_1 fixed. Thus, see for instance [4, pp. 4–5], F is holomorphic in the classical sense of two complex variables. This implies immediately many quite useful properties of F, in particular, it is of class $C^\infty(\Omega)$.

In one complex variable there exist two mutually conjugate Cauchy–Riemann operators, which characterize the usual holomorphy. Since in $\mathbb{BC}$ there exist three conjugations, it turns out that there exist four bicomplex operators, which characterize in a similar fashion the $\mathbb{BC}$-holomorphy. They are:

$$\frac{\partial}{\partial Z} := \frac{1}{2}\left(\frac{\partial}{\partial z_1} - \mathbf{j}\frac{\partial}{\partial z_2}\right), \qquad \frac{\partial}{\partial Z^\dagger} := \frac{1}{2}\left(\frac{\partial}{\partial z_1} + \mathbf{j}\frac{\partial}{\partial z_2}\right),$$

$$\frac{\partial}{\partial \bar{Z}} := \frac{1}{2}\left(\frac{\partial}{\partial \bar{z}_1} - \mathbf{j}\frac{\partial}{\partial \bar{z}_2}\right), \qquad \frac{\partial}{\partial Z^*} := \frac{1}{2}\left(\frac{\partial}{\partial \bar{z}_1} + \mathbf{j}\frac{\partial}{\partial \bar{z}_2}\right). \tag{2.6}$$

Theorem 2.1.1. Given $F \in C^1(\Omega, \mathbb{BC})$, it is $\mathbb{BC}$-holomorphic if and only if

$$\frac{\partial F}{\partial Z^\dagger}(Z) = \frac{\partial F}{\partial \bar{Z}}(Z) = \frac{\partial F}{\partial Z^*}(Z) = 0 \tag{2.7}$$

holds on Ω. Moreover, when it is true one has: $F'(Z) = \dfrac{\partial F}{\partial Z}(Z)$.

Let us see now how the idempotent representation of bicomplex numbers becomes crucial in a deeper understanding of the nature of $\mathbb{BC}$-holomorphic functions. Take a bicomplex function $F : \Omega \subset \mathbb{BC} \to \mathbb{BC}$ with Ω being a domain.

We write all the bicomplex numbers involved in idempotent form, for instance,

$$Z = \beta_1 \mathbf{e} + \beta_2 \mathbf{e}^\dagger = (\ell_1 + \mathbf{i}m_1)\mathbf{e} + (\ell_2 + \mathbf{i}m_2)\mathbf{e}^\dagger \,,$$

$$F(Z) = G_1(Z)\mathbf{e} + G_2(Z)\mathbf{e}^\dagger \,,$$

$$H = \eta_1 \mathbf{e} + \eta_2 \mathbf{e}^\dagger = (u_1 + \mathbf{i}v_1)\mathbf{e} + (u_2 + \mathbf{i}v_2)\mathbf{e}^\dagger \,.$$

Let us introduce the sets

$$\Omega_1 := \{\beta_1 \mid \beta_1 \mathbf{e} + \beta_2 \mathbf{e}^\dagger \in \Omega\} \subset \mathbb{C}(\mathbf{i})$$

and

$$\Omega_2 := \{\beta_2 \mid \beta_1 \mathbf{e} + \beta_2 \mathbf{e}^\dagger \in \Omega\} \subset \mathbb{C}(\mathbf{i}).$$

It is direct to prove that Ω_1 and Ω_2 are domains in $\mathbb{C}(\mathbf{i})$.

Theorem 2.1.2. A bicomplex function $F = G_1\mathbf{e} + G_2\mathbf{e}^\dagger : \Omega \subset \mathbb{BC} \to \mathbb{BC}$ of class C^1 is $\mathbb{BC}$-holomorphic if and only if the following two conditions hold:

(I) The component G_1, seen as a $\mathbb{C}(\mathbf{i})$-valued function of two complex variables (β_1, β_2) is holomorphic; what is more, it does not depend on the variable β_2 and thus G_1 is a holomorphic function of the variable β_1.

(II) The component G_2, seen as a $\mathbb{C}(\mathbf{i})$-valued function of two complex variables (β_1, β_2) is holomorphic; what is more, it does not depend on the variable β_1 and thus G_2 is a holomorphic function of the variable β_2.

Corollary 2.1.3. Let F be a $\mathbb{BC}$-holomorphic function in Ω, then F is of the form $F(Z) = G_1(\beta_1)\mathbf{e} + G_2(\beta_2)\mathbf{e}^\dagger$ with $Z = \beta_1 \mathbf{e} + \beta_2 \mathbf{e}^\dagger \in \Omega$ and its derivative is given by

$$F'(Z) = \mathbf{e} \cdot G_1'(\beta_1) + \mathbf{e}^\dagger \cdot G_2'(\beta_2)\,,$$

or equivalently:

$$F'(z_1 + \mathbf{j}z_2) = \mathbf{e} \cdot G_1'(z_1 - \mathbf{i}z_2) + \mathbf{e}^\dagger \cdot G_2'(z_1 + \mathbf{i}z_2)\,.$$

2.2 Bicomplex Matrices

We will denote by $\mathbb{BC}^{m \times n}$ the set of $m \times n$ matrices with bicomplex entries. For any such matrix $A = (a_{\ell j}) \in \mathbb{BC}^{m \times n}$, it is possible to consider both its cartesian and idempotent representation, which are obtained by accordingly decomposing each of its entries so that, for example, we have

$$A = \mathscr{A}_{1,\mathbf{i}}\mathbf{e} + \mathscr{A}_{2,\mathbf{i}}\mathbf{e}^\dagger = \mathscr{A}_{1,\mathbf{j}}\mathbf{e} + \mathscr{A}_{2,\mathbf{j}}\mathbf{e}^\dagger$$

where $\mathscr{A}_{1,\mathbf{i}}, \mathscr{A}_{2,\mathbf{i}} \in \mathbb{C}^{m \times n}(\mathbf{i})$ and $\mathscr{A}_{1,\mathbf{j}}, \mathscr{A}_{2,\mathbf{j}} \in \mathbb{C}^{m \times n}(\mathbf{j})$.

Of course the set $\mathbb{BC}^{m \times n}$ inherits many structures from $\mathbb{BC}$. It is obviously a $\mathbb{BC}$-module (a concept that will be discussed in detail in the next chapter) such that

$$\mathbb{BC}^{m \times n} = \mathbb{C}^{m \times n}(\mathbf{i}) \cdot \mathbf{e} + \mathbb{C}^{m \times n}(\mathbf{i}) \cdot \mathbf{e}^\dagger$$

$$= \mathbb{C}^{m \times n}(\mathbf{j}) \cdot \mathbf{e} + \mathbb{C}^{m \times n}(\mathbf{j}) \cdot \mathbf{e}^\dagger ,$$

where the summands are $\mathbb{BC}$-submodules of $\mathbb{BC}^{m \times n}$. In particular, given $\mathscr{B} \in \mathbb{C}^{m \times n}(\mathbf{i})$ then $\mathscr{B} \cdot \mathbf{e} \in \mathbb{C}^{m \times n}(\mathbf{i}) \cdot \mathbf{e}$ and $\mathbf{e}^\dagger \cdot (\mathscr{B} \cdot \mathbf{e}) = 0_{m \times n}$.

As in the scalar case, the operations over the matrices in the idempotent decomposition can be realized component-wise (though keeping in mind the non-commutativity of matrix multiplication).

Proposition 2.2.1. (See Theorem 3.1 in [5].) Let A be an $n \times n$ bicomplex matrix

$$A = \mathscr{A}_{1,\mathbf{i}}\mathbf{e} + \mathscr{A}_{2,\mathbf{i}}\mathbf{e}^\dagger = \mathscr{A}_{1,\mathbf{j}}\mathbf{e} + \mathscr{A}_{2,\mathbf{j}}\mathbf{e}^\dagger.$$

Then its determinant is given by

$$\det A = \det\mathscr{A}_{1,\mathbf{i}}\,\mathbf{e} + \det\mathscr{A}_{2,\mathbf{i}}\,\mathbf{e}^\dagger = \det\mathscr{A}_{1,\mathbf{j}}\,\mathbf{e} + \det\mathscr{A}_{2,\mathbf{j}}\,\mathbf{e}^\dagger.$$

Proof:
The proof can be done by induction on n. In the case $n = 2$ this is immediately demonstrated by the following easy calculation:

$$\det A = \det \begin{pmatrix} a_{11} & a_{12} \\ a_{21} & a_{22} \end{pmatrix} = a_{11}a_{22} - a_{12}a_{21}$$

$$= \left(a'_{11}\mathbf{e} + a''_{11}\mathbf{e}^\dagger\right)\left(a'_{22}\mathbf{e} + a''_{22}\mathbf{e}^\dagger\right)$$

$$- \left(a'_{12}\mathbf{e} + a''_{12}\mathbf{e}^\dagger\right)\left(a'_{21}\mathbf{e} + a''_{21}\mathbf{e}^\dagger\right)$$

$$= \left(a'_{11}a'_{22} - a'_{12}a'_{21}\right)\mathbf{e} + \left(a''_{11}a''_{22} - a''_{12}a''_{21}\right)\mathbf{e}^\dagger$$

$$= \det\mathscr{A}_1\mathbf{e} + \det\mathscr{A}_2\mathbf{e}^\dagger,$$

where the idempotent decompositions may be taken with coefficients either in $\mathbb{C}(\mathbf{i})$ or in $\mathbb{C}(\mathbf{j})$. For the general case, one can simply use the Laplace theorem, that gives the general formula for the determinant of an $n \times n$ matrix in terms of the determinants of suitable $(n-1) \times (n-1)$ matrices. $\qquad\square$

An immediate consequence of this result is the Binet theorem for bicomplex matrices; compare with Remark after Theorem 3.1 in [5].

Corollary 2.2.2. Let A and B be two square bicomplex matrices. Then

$$\det(AB) = \det A \cdot \det B.$$

Proof:

By using the previous result, we see that

$$\det(AB) = \det\left(\left(\mathscr{A}_{1,\mathbf{i}}\,\mathbf{e} + \mathscr{A}_{2,\mathbf{i}}\,\mathbf{e}^\dagger\right)\left(\mathscr{B}_{1,\mathbf{i}}\,\mathbf{e} + \mathscr{B}_{2,\mathbf{i}}\,\mathbf{e}^\dagger\right)\right)$$

$$= \det\left(\mathscr{A}_{1,\mathbf{i}}\mathscr{B}_{1,\mathbf{i}}\,\mathbf{e} + \mathscr{A}_{2,\mathbf{i}}\mathscr{B}_{2,\mathbf{i}}\,\mathbf{e}^\dagger\right)$$

$$= \det\left(\mathscr{A}_{1,\mathbf{i}}\mathscr{B}_{1,\mathbf{i}}\right)\mathbf{e} + \det\left(\mathscr{A}_{2,\mathbf{i}}\mathscr{B}_{2,\mathbf{i}}\right)\mathbf{e}^\dagger$$

$$= \det\mathscr{A}_{1,\mathbf{i}} \cdot \det\mathscr{B}_{1,\mathbf{i}}\,\mathbf{e} + \det\mathscr{A}_{2,\mathbf{i}} \cdot \det\mathscr{B}_{2,\mathbf{i}}\,\mathbf{e}^\dagger$$

$$= \left(\det\mathscr{A}_{1,\mathbf{i}}\,\mathbf{e} + \det\mathscr{A}_{2,\mathbf{i}}\,\mathbf{e}^\dagger\right)\left(\det\mathscr{B}_{1,\mathbf{i}}\,\mathbf{e} + \det\mathscr{B}_{2,\mathbf{i}}\,\mathbf{e}^\dagger\right)$$

$$= \det A \cdot \det B,$$

where one can take $\mathscr{A}_{1,\mathbf{j}}$, $\mathscr{A}_{2,\mathbf{j}}$, $\mathscr{B}_{1,\mathbf{j}}$, $\mathscr{B}_{2,\mathbf{j}}$ instead of their $\mathbb{C}(\mathbf{i})$-counterparts. This concludes the proof. $\qquad\square$

Analogously, one can use the idempotent representation of a bicomplex matrix to determine its invertibility.

Proposition 2.2.3. Let $A = \mathscr{A}_{1,\mathbf{i}}\,\mathbf{e} + \mathscr{A}_{2,\mathbf{i}}\,\mathbf{e}^\dagger = \mathscr{A}_{1,\mathbf{j}}\,\mathbf{e} + \mathscr{A}_{2,\mathbf{j}}\,\mathbf{e}^\dagger \in \mathbb{BC}^{n\times n}$, $\mathscr{A}_{1,\mathbf{i}}$, $\mathscr{A}_{2,\mathbf{i}} \in \mathbb{C}^{n\times n}(\mathbf{i})$, $\mathscr{A}_{1,\mathbf{j}}$, $\mathscr{A}_{2,\mathbf{j}} \in \mathbb{C}^{n\times n}(\mathbf{j})$ be a bicomplex matrix. Then A is invertible if and only if $\mathscr{A}_{1,\mathbf{i}}$ and $\mathscr{A}_{2,\mathbf{i}}$ are invertible in $\mathbb{C}^{n\times n}(\mathbf{i})$ and $\mathscr{A}_{1,\mathbf{j}}$, $\mathscr{A}_{2,\mathbf{j}}$ are invertible in $\mathbb{C}^{n\times n}(\mathbf{j})$.

Proof:

A matrix A is invertible if and only if there exists $B = \mathscr{B}_{1,\mathbf{i}}\,\mathbf{e} + \mathscr{B}_{2,\mathbf{i}}\,\mathbf{e}^\dagger \in \mathbb{BC}^{n\times n}$ such that $AB = BA = I$. This is equivalent to

$$I_{\mathbb{C}^{n \times n}(\mathrm{i})}\,\mathbf{e} + I_{\mathbb{C}^{n \times n}(\mathrm{i})}\,\mathbf{e}^\dagger = I_{\mathbb{BC}^{n \times n}} = \mathscr{A}_{1,\mathrm{i}}\,\mathscr{B}_{1,\mathrm{i}}\,\mathbf{e} + \mathscr{A}_{2,\mathrm{i}}\,\mathscr{B}_{2,\mathrm{i}}\,\mathbf{e}^\dagger$$

$$= \mathscr{B}_{1,\mathrm{i}}\,\mathscr{A}_{1,\mathrm{i}}\,\mathbf{e} + \mathscr{B}_{2,\mathrm{i}}\,\mathscr{A}_{2,\mathrm{i}}\,\mathbf{e}^\dagger$$

which is equivalent to $\mathscr{A}_{1,\mathrm{i}}\,\mathscr{B}_{1,\mathrm{i}} = I_{\mathbb{C}^{n \times n}(\mathrm{i})}$ and $\mathscr{A}_{2,\mathrm{i}}\,\mathscr{B}_{2,\mathrm{i}} = I_{\mathbb{C}^{n \times n}(\mathrm{i})}$. The same when one considers the invertibility of $\mathscr{A}_{1,\mathrm{j}}$ and $\mathscr{A}_{2,\mathrm{j}}$. □

The following result is immediate.

Corollary 2.2.4. A matrix $A \in \mathbb{BC}^{n \times n}$ is invertible if and only if $\det A \notin \mathfrak{S} \cup \{0\}$.

One can compare Proposition 2.2.3 and Corollary 2.2.4 with Theorem 3.3 in [5].

We can naturally introduce three conjugations on each bicomplex matrix $A = (a_{\ell j}) \in \mathbb{BC}^{m \times n}$, as follows:

$$A^\dagger := \left(a_{\ell j}^\dagger\right), \qquad \overline{A} := \left(\overline{a}_{\ell j}\right), \qquad A^* := \left(a_{\ell j}^*\right).$$

It is immediate to prove that all the conjugations are multiplicative operations, that is, each of them applied to the product of two matrices becomes the product of the conjugate matrices. Obviously these conjugations are additive operations also.

As usual, A^t denotes the transposed matrix:

$$A^t = (a_{j\ell}), \quad 1 \le j \le n, \quad 1 \le \ell \le m,$$

and we correspondingly have three adjoint matrices:

$$A^{t\dagger} := \left(A^t\right)^\dagger = \left(A^\dagger\right)^t = \left(a_{j\ell}^\dagger\right);$$

$$A^{t\,bar} := \overline{(A^t)} = \left(\overline{A}\right)^t = \left(\overline{a}_{j\ell}\right);$$

$$A^{t*} := \left(A^t\right)^* = \left(A^*\right)^t = \left(a_{j\ell}^*\right).$$

Since for any two compatible matrices one has that $(AB)^t = B^t A^t$, then $(AB)^{t\dagger} = B^{t\dagger} A^{t\dagger}$, $(AB)^{t\,bar} = B^{t\,bar} A^{t\,bar}$, $(AB)^{t*} = B^{t*} A^{t*}$.

The idempotent representations

$$A = \mathscr{A}_{1,\mathrm{i}}\,\mathbf{e} + \mathscr{A}_{2,\mathrm{i}}\,\mathbf{e}^\dagger = \mathscr{A}_{1,\mathrm{j}}\,\mathbf{e} + \mathscr{A}_{2,\mathrm{j}}\,\mathbf{e}^\dagger$$

give:

$$A^\dagger = \mathscr{A}_{2,\mathrm{i}}\,\mathbf{e} + \mathscr{A}_{1,\mathrm{i}}\,\mathbf{e}^\dagger = \mathscr{A}_{2,\mathrm{j}}^*\,\mathbf{e} + \mathscr{A}_{1,\mathrm{j}}^*\,\mathbf{e}^\dagger,$$

$$\overline{A} = \overline{\mathscr{A}}_{2,\mathrm{i}}\,\mathbf{e} + \overline{\mathscr{A}}_{1,\mathrm{i}}\,\mathbf{e}^\dagger = \mathscr{A}_{2,\mathrm{j}}\,\mathbf{e} + \mathscr{A}_{1,\mathrm{j}}\,\mathbf{e}^\dagger,$$

$$A^* = \overline{\mathscr{A}}_{1,\mathbf{i}}\,\mathbf{e} + \overline{\mathscr{A}}_{2,\mathbf{i}}\,\mathbf{e}^\dagger = \mathscr{A}_{1,\mathbf{j}}^*\,\mathbf{e} + \mathscr{A}_{2,\mathbf{j}}^*\,\mathbf{e}^\dagger,$$
$$A^t = \mathscr{A}_{1,\mathbf{i}}^t\,\mathbf{e} + \mathscr{A}_{2,\mathbf{i}}^t\,\mathbf{e}^\dagger = \mathscr{A}_{1,\mathbf{j}}^t\,\mathbf{e} + \mathscr{A}_{2,\mathbf{j}}^t\,\mathbf{e}^\dagger,$$

and hence

$$A^{t\,\dagger} = \mathscr{A}_{2,\mathbf{i}}^t\,\mathbf{e} + \mathscr{A}_{1,\mathbf{i}}^t\,\mathbf{e}^\dagger = \mathscr{A}_{2,\mathbf{j}}^{t\,*}\,\mathbf{e} + \mathscr{A}_{1,\mathbf{j}}^{t\,*}\,\mathbf{e}^\dagger,$$
$$A^{t\,bar} = \overline{\mathscr{A}}_{2,\mathbf{i}}^t\,\mathbf{e} + \overline{\mathscr{A}}_{1,\mathbf{i}}^t\,\mathbf{e}^\dagger = \mathscr{A}_{2,\mathbf{j}}^t\,\mathbf{e} + \mathscr{A}_{1,\mathbf{j}}^t\,\mathbf{e}^\dagger,$$
$$A^{t\,*} = \overline{\mathscr{A}}_{1,\mathbf{i}}^t\,\mathbf{e} + \overline{\mathscr{A}}_{2,\mathbf{i}}^t\,\mathbf{e}^\dagger = \mathscr{A}_{1,\mathbf{j}}^{t\,*}\,\mathbf{e} + \mathscr{A}_{2,\mathbf{j}}^{t\,*}\,\mathbf{e}^\dagger.$$

Similarly, we can define the notion of self-adjoint matrix with respect to each one of the conjugations introduced above; specifically, we will say that a matrix A is self-adjoint if, respectively, it satisfies one of the following equalities:

$$A = A^{t\,\dagger}, \quad A = A^{t\,bar}, \quad A = A^{t\,*}.$$

We can express these definitions in terms of idempotent components.

(a) The matrix A is $\dagger$-self-adjoint, or $\dagger$-Hermitian, if and only if

$$\mathscr{A}_{1,\mathbf{i}} = \mathscr{A}_{2,\mathbf{i}}^t, \quad \text{if and only if} \quad \mathscr{A}_{1,\mathbf{j}} = \mathscr{A}_{2,\mathbf{j}}^{t\,*},$$

which is true if and only if

$$A^t = A^\dagger;$$

thus all $\dagger$-self-adjoint matrices are of the form

$$A = \mathscr{A}_{1,\mathbf{i}}\,\mathbf{e} + \mathscr{A}_{1,\mathbf{i}}^t\,\mathbf{e}^\dagger = \mathscr{A}_{1,\mathbf{j}}\,\mathbf{e} + \mathscr{A}_{1,\mathbf{j}}^{t\,*}\,\mathbf{e}^\dagger,$$

with $\mathscr{A}_{1,\mathbf{i}}$ an arbitrary matrix in $\mathbb{C}^{n\times n}(\mathbf{i})$, and $\mathscr{A}_{1,\mathbf{j}}$ any matrix in $\mathbb{C}^{n\times n}(\mathbf{j})$. Note that $\mathscr{A}_{1,\mathbf{j}}^{t\,*}$ is the usual $\mathbb{C}(\mathbf{j})$ adjoint of $\mathscr{A}_{1,\mathbf{j}}$.

(b) The matrix A is bar-self-adjoint, or bar-Hermitian, if and only if

$$\mathscr{A}_{1,\mathbf{i}} = \overline{\mathscr{A}}_{2,\mathbf{i}}^t, \quad \text{if and only if} \quad \mathscr{A}_{1,\mathbf{j}} = \mathscr{A}_{2,\mathbf{j}}^t,$$

which is true if and only if

$$A^t = \overline{A};$$

thus all bar-self-adjoint matrices are of the form

$$A = \mathscr{A}_{1,\mathbf{i}}\,\mathbf{e} + \overline{\mathscr{A}}_{1,\mathbf{i}}^t\,\mathbf{e}^\dagger = \mathscr{A}_{1,\mathbf{j}}\,\mathbf{e} + \mathscr{A}_{1,\mathbf{j}}^t\,\mathbf{e}^\dagger,$$

with $\mathscr{A}_{1,\mathbf{i}}$ an arbitrary matrix in $\mathbb{C}^{n\times n}(\mathbf{i})$, and $\mathscr{A}_{1,\mathbf{j}}$ any matrix in $\mathbb{C}^{n\times n}(\mathbf{j})$. Note that $\overline{\mathscr{A}}_{1,\mathbf{i}}^{t}$ is the usual $\mathbb{C}(\mathbf{i})$ adjoint of $\mathscr{A}_{1,\mathbf{i}}$.

(c) The matrix A is $*$-self-adjoint, or $*$-Hermitian, if and only if

$$\mathscr{A}_{1,\mathbf{i}} = \overline{\mathscr{A}}_{1,\mathbf{i}}^{t}, \quad \mathscr{A}_{2,\mathbf{i}} = \overline{\mathscr{A}}_{2,\mathbf{i}}^{t},$$

if and only if

$$\mathscr{A}_{1,\mathbf{j}} = \mathscr{A}_{1,\mathbf{j}}^{t\,*}, \quad \mathscr{A}_{2,\mathbf{j}} = \mathscr{A}_{2,\mathbf{i}}^{t\,*};$$

that is, both $\mathscr{A}_{1,\mathbf{i}}$ and $\mathscr{A}_{2,\mathbf{i}}$ are usual $\mathbb{C}(\mathbf{i})$ self-adjoint matrices, and both $\mathscr{A}_{1,\mathbf{j}}$, $\mathscr{A}_{2,\mathbf{j}}$ are usual $\mathbb{C}(\mathbf{j})$ self-adjoint matrices.

In what follows we are interested in $*$-self-adjointness since this property implies a form of hyperbolic "positiveness" of bicomplex matrices which we will find quite useful.

Definition 2.2.6. A $*$-self-adjoint matrix $A \in \mathbb{BC}^{n\times n}$ is called *hyperbolic positive*, if for every column $c \in \mathbb{BC}^{n}$,

$$c^{*t} \cdot A \cdot c \in \mathbb{D}^{+}. \tag{2.8}$$

In this case we write $A \succ 0$. Given two bicomplex matrices A, B, we say that $A \succ B$ if and only if $A - B \succ 0$.

Proposition 2.2.7. Let

$$A = A_1 + \mathbf{j}A_2 = \mathscr{A}_1 \mathbf{e} + \mathscr{A}_2 \mathbf{e}^{\dagger}, \tag{2.9}$$

be an element of $\mathbb{BC}^{n}$, with A_1, A_2, $\mathscr{A}_1$ and $\mathscr{A}_2$ in $\mathbb{C}^{n\times n}(\mathbf{i})$. Then, the following are equivalent:

(a) $A \succ 0$.
(b) Both $\mathscr{A}_1$ and $\mathscr{A}_2$ are complex positive matrices.
(c) $A_1 \geq 0$, the matrix A_2 is skew self adjoint, that is, $A_2 + \overline{A}_2^{t} = 0$, and

$$- A_1 \leq \mathbf{i}A_2 \leq A_1. \tag{2.10}$$

Proof:
Assume (a). Take a column $c \in \mathbb{BC}^{n}$ with representations

$$c = c_1 + \mathbf{j}c_2 = \zeta_1 \mathbf{e} + \zeta_2 \mathbf{e}^{\dagger},$$

where the various columns are in $\mathbb{C}^{n}(\mathbf{i})$. Then,

$$c^{*t} Ac = \overline{\zeta}_1^{t} \mathscr{A}_1 \zeta_1 \mathbf{e} + \overline{\zeta}_2^{t} \mathscr{A}_2 \zeta_2 \mathbf{e}^{\dagger}. \tag{2.11}$$

Thus, by definition, $A > 0$ implies that both $\mathscr{A}_1$ and $\mathscr{A}_2$ are $\mathbb{C}(\mathbf{i})$ positive, and (b) holds.

Assume (b). Since

$$\mathscr{A}_1 = A_1 - \mathbf{i}A_2 \quad \text{and} \quad \mathscr{A}_2 = A_1 + \mathbf{i}A_2,$$

then

$$A_1 = \frac{1}{2}(\mathscr{A}_1 + \mathscr{A}_2) \geq 0;$$

moreover, $\mathbf{i}A_2 = A_1 - \mathscr{A}_1$ is self-adjoint, and hence A_2 is skew self-adjoint. Furthermore, still in view of (b),

$$A_1 - \mathbf{i}A_2 \geq 0 \quad \text{and} \quad A_1 + \mathbf{i}A_2 \geq 0,$$

and thus we obtain (2.10) and (c) holds. Finally when (c) holds, both the matrices $\mathscr{A}_1$ and $\mathscr{A}_2$ are positive, and thus (a) holds as well in view of (2.11). □

Proposition 2.2.8. Let $A \in \mathbb{BC}^{n \times n}$. The following are equivalent:

(1) A is hyperbolic positive.
(2) $A = B^{*t} \cdot B$ where $B \in \mathbb{BC}^{m \times n}$ for some $m \in \mathbb{N}$.
(3) $A = C^2$ where the matrix C is hyperbolic positive.

Proof:
Let $A \in \mathbb{BC}^{n \times n}$ be represented as in (2.9), and assume that (1) holds. Then, by the preceding theorem, we have $\mathscr{A}_1 \geq 0$ and $\mathscr{A}_2 \geq 0$, and thus we can write

$$\mathscr{A}_1 = \overline{U}^t U \quad \text{and} \quad \mathscr{A}_2 = \overline{V}^t V, \tag{2.12}$$

where U and V are matrices in $\mathbb{C}^{n \times n}(\mathbf{i})$. Thus $A = B^{*t}B$ with $B := U\mathbf{e} + V\mathbf{e}^{\dagger}$, so that (2) holds with $m = n$. Assume now that (2) holds with $B \in \mathbb{BC}^{m \times n}$ for some $m \in \mathbb{N}$. Writing $B = U\mathbf{e} + V\mathbf{e}^{\dagger}$, where now U and V belong to $\mathbb{C}^{m \times n}(\mathbf{i})$ we have

$$A = \overline{U}^t U\mathbf{e} + \overline{V}^t V\mathbf{e}^{\dagger},$$

and so (1) holds. The equivalence with (3) stems from the fact that U and V in (2.12) can be chosen positive. □

In the case of a complex matrix, it is equivalent to say that a positive matrix A is Hermitian and to say that its eigenvalues are positive. We now give the corresponding result in the setting of $\mathbb{BC}$. We note that the existence of zero divisors creates problems in an attempt to classify eigenvalues in general. For instance, take two non zero elements a and b such that $ab = 0$. Then every $c \in \mathbb{BC}$ such that $bc = 0$ is an eigenvalue of the matrix

$$\begin{pmatrix} a & a \\ a & a \end{pmatrix} \quad \text{with eigenvector} \quad \begin{pmatrix} b \\ b \end{pmatrix},$$

since

$$\begin{pmatrix} a & a \\ a & a \end{pmatrix} \begin{pmatrix} b \\ b \end{pmatrix} = c \begin{pmatrix} b \\ b \end{pmatrix} = \begin{pmatrix} 0 \\ 0 \end{pmatrix}.$$

There is however a relation between the eigenvalues and eigenvectors of a bicomplex matrix A and those of its idempotent components $\mathscr{A}_1$ and $\mathscr{A}_2$. Indeed, let $\lambda = \gamma_1 \, \mathbf{e} + \gamma_2 \, \mathbf{e}^\dagger \in \mathbb{BC} \setminus \{0\}$ be an eigenvalue of A with a corresponding eigenvector $u = v_1 \, \mathbf{e} + v_2 \, \mathbf{e}^\dagger$, then

$$Au = \lambda u,$$

which is equivalent to

$$\begin{cases} \mathscr{A}_1 \, v_1 = \gamma_1 \, v_1, \\ \\ \mathscr{A}_2 \, v_2 = \gamma_2 \, v_2. \end{cases}$$

If λ is not a zero divisor and $v_1 \neq 0$, $v_2 \neq 0$ then λ is an eigenvalue of A if and only if γ_1 is an eigenvalue of $\mathscr{A}_1$ and γ_2 is an eigenvalue of $\mathscr{A}_2$.

Theorem 2.2.9. A matrix $A \in \mathbb{BC}^{n \times n}$ is hyperbolic positive if and only if

1. A is $*$-Hermitian;
2. none of its eigenvalues is a zero divisor in $\mathbb{D}^+$.

Proof:
That A is hyperbolic positive is equivalent to the complex matrices $\mathscr{A}_1$, $\mathscr{A}_2$ being positive, which is in turn equivalent to stating that $\mathscr{A}_1$ and $\mathscr{A}_2$ are complex Hermitian and any of their eigenvalues are a positive number. Finally, this is equivalent to say that A is $*$-Hermitian and that none of its eigenvalues is a zero divisor in $\mathbb{D}^+$. □

Corollary 2.2.10. A matrix $A \in \mathbb{BC}^{n \times n}$ is hyperbolic positive if and only if:

1. A is $*$-Hermitian;
2. if $\lambda = \lambda_1 + \mathbf{j} \lambda_2$ is an eigenvalue for A, then $\lambda_1 > 0$, $\mathbf{i} \lambda_2 \in \mathbb{R}$ and

$$-\lambda_1 < \mathbf{i} \lambda_2 < \lambda_1 \,;$$

Proof:
This is because $\lambda = \lambda_1 + \mathbf{j} \lambda_2 = \gamma_1 \, \mathbf{e} + \gamma_2 \, \mathbf{e}^\dagger$ with $\gamma_1 = \lambda_1 - \mathbf{i} \lambda_2$ and $\gamma_2 = \lambda_1 + \mathbf{i} \lambda_2$. □

Remark 2.2.11. The last inequality in the statement of this corollary is equivalent to the system

$$\begin{cases} \lambda_1 - \mathbf{i} \lambda_2 > 0, \\ \\ \lambda_1 + \mathbf{i} \lambda_2 > 0. \end{cases}$$

It is known that in the case of complex matrices every eigenvector corresponds to only one eigenvalue. This is not the case for bicomplex matrices. Specifically, a bicomplex eigenvector can correspond to an infinite family of bicomplex eigenvalues.

We restrict our considerations to hyperbolic positive matrices. Let A be such a matrix, and let $\lambda = \gamma_1\,\mathbf{e} + \gamma_2\,\mathbf{e}^{\dagger}$ be one of its eigenvalues (in particular λ is a non zero divisor in $\mathbb{D}^{+}$). First of all let us show that any such eigenvalue has an eigenvector of the form $u = v_2\,\mathbf{e}^{\dagger}$ with $v_2 \in \mathbb{C}^n(\mathbf{i})$. Since γ_2 is an eigenvalue of $\mathscr{A}_2$, let v_2 be a corresponding eigenvector: $\mathscr{A}_2\,v_2 = \gamma_2\,v_2$. Consider $u := v_2\,\mathbf{e}^{\dagger}$. Let us show that it is an eigenvector of A corresponding to the above λ. Indeed, $A\,u = \mathscr{A}_2\,v_2\,\mathbf{e}^{\dagger} = \gamma_2\,v_2\,\mathbf{e}^{\dagger}$ and $\lambda\,u = \gamma_2\,v_2\,\mathbf{e}^{\dagger}$, thus $A\,u = \lambda\,u$.

Now we are in a position to show that this eigenvector corresponds to an infinite family of eigenvalues. For any $r > 0$ set $\lambda_r := r\,\mathbf{e} + \gamma_2\,\mathbf{e}^{\dagger}$, then $\lambda_r\,u = \gamma_2\,v_2\,\mathbf{e}^{\dagger} = A\,u$. Hence the whole family $\{\,\lambda_r \mid r > 0\,\}$ consists of the eigenvalues of A with the same eigenvector $u \in \mathbb{BC}^n_{\mathbf{e}^{\dagger}}$.

We now study bicomplex $*$-unitary matrices, that is, matrices $U \in \mathbb{BC}^{n \times n}$ such that $U\,U^{*t} = U^{*t}U = I_n$.

Proposition 2.2.12. Let $U = U_1 + \mathbf{j}\,U_2 = \mathscr{U}_1\,\mathbf{e} + \mathscr{U}_2\,\mathbf{e}^{\dagger} \in \mathbb{BC}^{n \times n}$. Then U is unitary if and only if its idempotent components are complex unitary matrices or, equivalently, its cartesian components satisfy

$$U_1\,\overline{U}_1^{t} + U_2\,\overline{U}_2^{t} = \overline{U}_1^{t}\,U_1 + \overline{U}_2^{t}\,U_2 = I_n$$

and

$$U_2\,\overline{U}_1^{t} = U_1\,\overline{U}_2^{t}, \qquad \overline{U}_1^{t}\,U_2 = \overline{U}_2^{t}\,U_1.$$

Proof:
Since $U^{*t} = \overline{U}_1^{t} - \mathbf{j}\,\overline{U}_2^{t} = \overline{\mathscr{U}}_1^{t}\,e + \overline{\mathscr{U}}_2^{t}\,\mathbf{e}^{\dagger}$, then for the idempotent representation we have:

$$\mathscr{U}_1\,\overline{\mathscr{U}}_1^{t}\,\mathbf{e} + \mathscr{U}_2\,\overline{\mathscr{U}}_2^{t}\,\mathbf{e}^{\dagger} = \mathbf{e}\,I_n + \mathbf{e}^{\dagger}\,I_n$$

and

$$\overline{\mathscr{U}}_1^{t}\,\mathscr{U}_1\,\mathbf{e} + \overline{\mathscr{U}}_2^{t}\,\mathscr{U}_2\,\mathbf{e}^{\dagger} = \mathbf{e}\,I_n + \mathbf{e}^{\dagger}\,I_n.$$

This means that $\mathscr{U}_1$ and $\mathscr{U}_2$ are complex unitary matrices.

Similarly for the cartesian representation we have:

$$U_1\,\overline{U}_1^{t} + U_2\,\overline{U}_2^{t} + \mathbf{j}\left(U_2\,\overline{U}_1^{t} - U_1\,\overline{U}_2^{t}\right) = I_n,$$

$$\overline{U}_1^{t}\,U_1 + \overline{U}_2^{t}\,U_2 + \mathbf{j}\left(\overline{U}_1^{t}\,U_2 - \overline{U}_2^{t}\,U_1\right) = I_n,$$

which means that

$$U_1 \overline{U}_1^t + U_2 \overline{U}_2^t = \overline{U}_1^t U_1 + \overline{U}_2^t U_2 = I_n$$

and

$$U_2 \overline{U}_1^t = U_1 \overline{U}_2^t, \qquad \overline{U}_1^t U_2 = \overline{U}_2^t U_1.$$

The result follows.   □

References

1. G.B. Price, *An Introduction to Multicomplex Spaces and Functions* (Marcel Dekker, New York, 1991)
2. M.E. Luna-Elizarrarás, M. Shapiro, D.C. Struppa, A. Vajiac, Complex Laplacian and derivatives of bicomplex functions. Complex Anal. Oper. Theory **7**(5), 1675–1711 (2013)
3. M.E. Luna-Elizarrarás, M. Shapiro, D.C. Struppa, A. Vajiac, Bicomplex numbers and their holomorphic functions. In preparation
4. S.G. Krantz, *Function Theory of Several Complex Variables* (AMS Chelsea Publishing, Providence, Rhode Island, 1992)
5. R. Gervais Lavoie, L. Marchildon, D. Rochon, Finite-dimensional bicomplex Hilbert spaces. Adv. Appl. Clifford Algebras **21**(3), 561–581 (2011)

Chapter 3
$\mathbb{BC}$-Modules

Abstract The chapter deals with the definition of a bicomplex module and some structures induced on it by the properties of bicomplex scalars. It is shown how one can construct a bicomplex module beginning with a pair of complex linear spaces.

Keywords Bicomplex modules · Involutions · Idempotent decomposition of a bicomplex module

3.1 $\mathbb{BC}$-Modules and Involutions on them

Let X be a $\mathbb{BC}$-module. It turns out that some structural peculiarities of the set $\mathbb{BC}$ are immediately manifested in any bicomplex module, in contrast to the cases of real, complex, or even quaternionic linear spaces where the structure of linear space does not imply anything immediate about the space itself.

Indeed, consider the sets

$$X_{\mathbf{e}} := \mathbf{e} \cdot X \quad \text{and} \quad X_{\mathbf{e}^\dagger} := \mathbf{e}^\dagger \cdot X.$$

Since

$$X_{\mathbf{e}} \cap X_{\mathbf{e}^\dagger} = \{0\},$$

and

$$X = X_{\mathbf{e}} + X_{\mathbf{e}^\dagger}, \tag{3.1}$$

one can define two mappings

$$P : X \to X, \quad Q : X \to X$$

by

D. Alpay et al., *Basics of Functional Analysis with Bicomplex Scalars, and Bicomplex Schur Analysis*, SpringerBriefs in Mathematics, DOI: 10.1007/978-3-319-05110-9_3, © The Author(s) 2014

$$P(x) := \mathbf{e}\,x \qquad Q(x) := \mathbf{e}^\dagger x\,.$$

Since

$$P + Q = Id_X, \quad P \circ Q = Q \circ P = 0, \quad P^2 = P \quad \text{and} \quad Q^2 = Q,$$

then the operators P and Q are mutually complementary projectors. Formula (3.1) is called the idempotent decomposition of X, and it plays an extremely important role in what follows. In particular, it allows to realize component-wise the operations on X: if $x = \mathbf{e}x + \mathbf{e}^\dagger x$, $y = \mathbf{e}y + \mathbf{e}^\dagger y$ and if $\lambda = \lambda_1\mathbf{e} + \lambda_2\mathbf{e}^\dagger$ then $x + y = (\mathbf{e}x + \mathbf{e}y) + (\mathbf{e}^\dagger x + \mathbf{e}^\dagger y)$, $\lambda x = \lambda_1 x\mathbf{e} + \lambda_2 x\mathbf{e}^\dagger$.

In what follows we will write $X_{\mathbb{C}(\mathbf{i})}$ or $X_{\mathbb{C}(\mathbf{j})}$ whenever X is considered as a $\mathbb{C}(\mathbf{i})$ or $\mathbb{C}(\mathbf{j})$ linear space respectively.

Since $X_\mathbf{e}$ and $X_{\mathbf{e}^\dagger}$ are $\mathbb{R}$-, $\mathbb{C}(\mathbf{i})$– and $\mathbb{C}(\mathbf{j})$–linear spaces as well as $\mathbb{BC}$-modules, we have that $X = X_\mathbf{e} \oplus X_{\mathbf{e}^\dagger}$ where the direct sum $\oplus$ can be understood in the sense of $\mathbb{R}$-, $\mathbb{C}(\mathbf{i})$– or $\mathbb{C}(\mathbf{j})$–linear spaces, as well as $\mathbb{BC}$-modules.

Note that given $w \in X_\mathbf{e}$ there exists $x \in X$ such that $w = P(x) = \mathbf{e}x$ which implies $\mathbf{e}\,w = \mathbf{e}^2 x = \mathbf{e}x = w$, i.e., any $w \in X_\mathbf{e}$ is such that $w = \mathbf{e}\,w$. Similarly if $t \in X_{\mathbf{e}^\dagger}$, then $t = \mathbf{e}^\dagger t$. As a matter of fact these identities characterize the elements in $X_\mathbf{e}$ and $X_{\mathbf{e}^\dagger}$: $w \in X_\mathbf{e}$ if and only if $w = \mathbf{e} \cdot w$ and $t \in X_{\mathbf{e}^\dagger}$ if and only if $t = \mathbf{e}^\dagger \cdot t$.

In order to be able to say more about $X_\mathbf{e}$ and $X_{\mathbf{e}^\dagger}$ we need additional assumptions about X. First of all, we are interested in the cartesian-like decompositions of $\mathbb{BC}$–modules connected with the analogs of the bicomplex conjugations.

Definition 3.1.2 An involution on a $\mathbb{BC}$-module X is a map

$$\mathcal{I}n : X \to X$$

such that

(1) $\mathcal{I}n(x + y) = \mathcal{I}n(x) + \mathcal{I}n(y)$ for any $x, y \in X$;
(2) $\mathcal{I}n^2 = I_X$, the identity map;
    and one of the following conditions holds for all $Z \in \mathbb{BC}$ and all $x \in X$:
(3a) $\mathcal{I}n(Z\,x) = \overline{Z}\,\mathcal{I}n_{bar}(x)$; in this case we will say this is a bar-involution, and we will indicate it by the symbol $\mathcal{I}n_{bar}$; in particular we note that such an involution is $\mathbb{C}(\mathbf{j})$-linear while it is $\mathbb{C}(\mathbf{i})$-antilinear and $\mathbb{D}$-antilinear;
(3b) $\mathcal{I}n(Z\,x) = Z^\dagger\,\mathcal{I}n_\dagger(x)$; in this case we will say this is a †-involution, and we will indicate it by the symbol $\mathcal{I}n_\dagger$; we note that such an involution is $\mathbb{C}(\mathbf{i})$-linear while it is $\mathbb{C}(\mathbf{j})$– and $\mathbb{D}$-antilinear;
(3c) $\mathcal{I}n(Z\,x) = Z^*\,\mathcal{I}n_*(x)$; in this case we will say this is a *-involution, and we will indicate it by the symbol $\mathcal{I}n_*$; we note that such an involution is $\mathbb{D}$-linear while it is $\mathbb{C}(\mathbf{j})$– and $\mathbb{C}(\mathbf{i})$-antilinear.

Each of these involutions induces a corresponding decomposition in X. Let us briefly describe them, beginning with $\mathcal{I}n_\dagger$. Define

$$X_{1,\dagger} := \{x \in X \mid \mathcal{I}n_\dagger \, x = x\} \quad \text{and} \quad X_{2,\dagger} := \{x \in X \mid \mathcal{I}n_\dagger \, x = -x\}.$$

We then have the following result:

Proposition 3.1.3 With the above notations it holds:

$$X = X_{1,\dagger} + X_{2,\dagger} = X_{1,\dagger} + \mathbf{j} \, X_{1,\dagger}. \tag{3.2}$$

Proof:
We first show that $\mathbf{j} X_{1,\dagger} = X_{2,\dagger}$. Indeed if $x \in X_{1,\dagger}$, then $\mathcal{I}n_\dagger(\mathbf{j} x) = -\mathbf{j} \mathcal{I}n_\dagger(x) = -\mathbf{j} x$, that is, $\mathbf{j} x \in X_{2,\dagger}$, which means that $\mathbf{j} X_{1,\dagger} \subset X_{2,\dagger}$. Conversely if $w \in X_{2,\dagger}$, then $\mathcal{I}n_\dagger(-\mathbf{j} w) = \mathbf{j} \mathcal{I}n_\dagger(w) = -\mathbf{j} w$. This implies $-\mathbf{j} w = x \in X_{1,\dagger}$ or, equivalently, $w = \mathbf{j} x$, thus

$$\mathbf{j} X_{1,\dagger} = X_{2,\dagger}.$$

Moreover, given $x \in X$, define

$$x_1 := \frac{1}{2}(x + \mathcal{I}n_\dagger \, x) \quad \text{and} \quad x_2 := \frac{1}{2}(x - \mathcal{I}n_\dagger \, x).$$

Clearly $x_1 \in X_{1,\dagger}$, $x_2 \in X_{2,\dagger}$, and $x = x_1 + x_2 = x_1 + \mathbf{j}(-\mathbf{j} x_2)$, proving the claim. $\square$

We will say that $X_{1,\dagger}$ is the set of $\mathbb{C}(\mathbf{i})$–elements in X since it is of course a $\mathbb{C}(\mathbf{i})$–linear subspace of X. The representation (3.2) is a generalization of the cartesian representation of $\mathbb{BC}$.

Note that any $\dagger$–involution on a $\mathbb{BC}$-module X is consistent with its idempotent representation. Specifically, if $x \in X$, $x = x_1 + \mathbf{j} y_1$, x_1, $y_1 \in X_{1,\dagger}$ we have that

$$x = \mathbf{e}(x_1 - \mathbf{i} y_1) + \mathbf{e}^\dagger(x_1 + \mathbf{i} y_1). \tag{3.3}$$

As we know the idempotent representation of a $\mathbb{BC}$-module always exists, but if the $\mathbb{BC}$-module has also a $\dagger$–involution which generates a cartesian decomposition, then both representations are related in the same way as it happens in $\mathbb{BC}$. Note that both elements $x_1 - \mathbf{i} y_1$ and $x_1 + \mathbf{i} y_1$ are $\mathbb{C}(\mathbf{i})$-elements, and in general we cannot say more.

Exactly the same analysis can be made for $\mathcal{I}n_{bar}$ and $\mathcal{I}n_*$. If we set

$$X_{1,bar} := \{x \in X \mid \mathcal{I}n_{bar}(x) = x\}$$

(the set of $\mathbb{C}(\mathbf{j})$-complex elements) and

$$X_{2,bar} := \{x \in X \mid \mathcal{I}n_{bar}(x) = -x\},$$

then, again, $X_{2,bar} = \mathbf{i} X_{1,bar}$ and

$$X = X_{1,bar} + X_{2,bar} = X_{1,bar} + \mathbf{i} X_{1,bar}. \tag{3.4}$$

This decomposition is also consistent with the idempotent one, since for $x = u_1 + \mathbf{i}\, u_2 \in X$ with u_1, $u_2 \in X_{1,bar}$ there holds:

$$x = \mathbf{e}\,(u_1 - \mathbf{j}\, u_2) + \mathbf{e}^\dagger\,(u_1 + \mathbf{j}\, u_2)\,. \tag{3.5}$$

If X has the two involutions $\mathcal{I}n_\dagger$ and $\mathcal{I}n_{bar}$, then both decompositions (3.3) and (3.5) hold for any $x \in X$; moreover

$$\mathbf{e}\,(u_1 - \mathbf{j}\, u_2) = \mathbf{e}\,(x_1 - \mathbf{i}\, y_1)$$

and

$$\mathbf{e}^\dagger\,(u_1 + \mathbf{j}\, u_2) = \mathbf{e}^\dagger\,(x_1 + \mathbf{i}\, y_1)$$

although not necessarily $u_1 - \mathbf{j}\, u_2 = x_1 - \mathbf{i}\, y_1$ and $u_1 + \mathbf{j}\, u_2 = x_1 + \mathbf{i}\, y_1$.

Finally, for $\mathcal{I}n_*$ we have:

$$X_{1,*} := \{\, x \in X \mid \mathcal{I}n_*(x) = x \,\}$$

(the set of hyperbolic elements) and

$$X_{2,*} := \{\, x \in X \mid \mathcal{I}n_*(x) = -x \,\}\,,$$

with $X_{2,*} = \mathbf{ij}\, X_{1,*}$, and

$$X = X_{1,*} + X_{2,*} = X_{1,*} + \mathbf{ij}\, X_{1,*}\,.$$

The existence of the idempotent representation does not necessarily imply the existence of an involution. Note first that a $\dagger$–involution exists if and only if the module X has a decomposition (3.2), hence, having only the hypothesis of the idempotent representation and trying to look for components x_1, x_2 such that $x = x_1 + \mathbf{j}\, x_2$, one is led to the system

$$2\,\mathbf{e}\, x = (1 + \mathbf{ij})\, x_1 + (\mathbf{j} - \mathbf{i})\, x_2\,;$$

$$2\,\mathbf{e}^\dagger\, x = (1 - \mathbf{ij})\, x_1 + (\mathbf{j} + \mathbf{i})\, x_2\,. \tag{3.6}$$

Since $\mathbf{e}$ and $\mathbf{e}^\dagger$ are zero divisors there is no guarantee that the system (3.6) has a solution.

3.2 Constructing a $\mathbb{BC}$-Module from Two Complex Linear Spaces

Let X_1, X_2 be $\mathbb{C}(\mathbf{i})$-linear spaces. We want to construct a $\mathbb{BC}$-module X such that, with suitable meaning of the symbols,

$$X = \mathbf{e}X_1 + \mathbf{e}^\dagger X_2. \tag{3.7}$$

This means that we have to give a precise meaning to the symbols $\mathbf{e}X_1$ and $\mathbf{e}^\dagger X_2$. For this purpose, consider the $\mathbb{C}(\mathbf{i})$-linear space $\mathbf{e}\,\mathbb{C}(\mathbf{i})$ which is a $\mathbb{C}(\mathbf{i})$-linear subspace of $\mathbb{BC}$, and define the tensor products

$$\mathbf{e}X_1 := \mathbf{e}\,\mathbb{C}(\mathbf{i}) \otimes_{\mathbb{C}(\mathbf{i})} X_1$$

and

$$\mathbf{e}^\dagger X_2 := \mathbf{e}^\dagger\,\mathbb{C}(\mathbf{i}) \otimes_{\mathbb{C}(\mathbf{i})} X_2\,.$$

It is clear that both $\mathbf{e}X_1$ and $\mathbf{e}^\dagger X_2$ are $\mathbb{C}(\mathbf{i})$-linear spaces.

From the definition of tensor product any elementary tensor in $\mathbf{e}X_1$ is of the form $\mathbf{e}\lambda \otimes x$ with $\lambda \in \mathbb{C}(\mathbf{i})$, hence

$$\mathbf{e}\lambda \otimes x = \mathbf{e} \otimes \lambda x = \mathbf{e} \otimes x_1.$$

That is, any element in $\mathbf{e}X_1$ is of the form $\mathbf{e} \otimes x_1$, with $x_1 \in X_1$, which we will write as $\mathbf{e}x_1$. Similarly with $\mathbf{e}^\dagger X_2$.

Consider now the cartesian product $\mathbf{e}X_1 \times \mathbf{e}^\dagger X_2$, where as usual, $\mathbf{e}X_1$ is seen as $\mathbf{e}X_1 \times \{0\}$ and $\mathbf{e}^\dagger X_2$ is seen as $\{0\} \times \mathbf{e}^\dagger X_2$. Since $\mathbf{e}X_1 \times \mathbf{e}^\dagger X_2$ is an additive abelian group we have endowed the sum $\mathbf{e}X_1 + \mathbf{e}^\dagger X_2$ with a meaning: for any $x_1 \in X_1$, $x_2 \in X_2$

$$\mathbf{e}x_1 + \mathbf{e}^\dagger x_2 := (\mathbf{e}x_1, \mathbf{e}^\dagger x_2) \in \mathbf{e}X_1 \times \mathbf{e}^\dagger X_2.$$

Hence the right-hand side in (3.7) is defined already but for the moment as a $\mathbb{C}(\mathbf{i})$-linear space only. Let us endow it with the structure of $\mathbb{BC}$-module. Given $\lambda = \beta_1\mathbf{e} + \beta_2\mathbf{e}^\dagger \in \mathbb{BC}$ and $\mathbf{e}x_1 + \mathbf{e}^\dagger x_2 \in \mathbf{e}X_1 + \mathbf{e}^\dagger X_2$, we set:

$$\lambda(\mathbf{e}x_1 + \mathbf{e}^\dagger x_2) = (\beta_1\mathbf{e} + \beta_2\mathbf{e}^\dagger) \cdot (\mathbf{e}x_1 + \mathbf{e}^\dagger x_2) := \mathbf{e}(\beta_1 x_1) + \mathbf{e}^\dagger(\beta_2 x_2).$$

It is immediate to check that this is a well-defined multiplication of the elements of $\mathbf{e}X_1 + \mathbf{e}^\dagger X_2$ by bicomplex scalars. Thus formula (3.7) defines indeed a $\mathbb{BC}$-module.

Of course, in complete accordance with Sect. 3.1, one has

$$X_\mathbf{e} := \mathbf{e}X = \mathbf{e}X_1;$$

$$X_{\mathbf{e}^\dagger} := \mathbf{e}^\dagger X = \mathbf{e}^\dagger X_2,$$

While X_1 and X_2 are initially $\mathbb{C}(\mathbf{i})$-linear spaces only, the new sets $\mathbf{e}X_1$ and $\mathbf{e}^\dagger X_2$ are $\mathbb{C}(\mathbf{j})$-linear spaces as well, and therefore they become $\mathbb{BC}$–modules. This is a consequence of the fact that since $\mathbf{je} = -\mathbf{ie}$ and $\mathbf{je}^\dagger = \mathbf{ie}^\dagger$, we can set

$$\mathbf{j}(\mathbf{e}x_1) := \mathbf{e}(-\mathbf{i}x_1) \in \mathbf{e}X_1, \quad \forall x_1 \in X_1;$$

$$\mathbf{j}(\mathbf{e}^\dagger x_2) := \mathbf{e}^\dagger(\mathbf{i}x_2) \in \mathbf{e}^\dagger X_2, \quad \forall x_2 \in X_2.$$

Note that X_1 and X_2 are not, in general, the "cartesian components" of a $\mathbb{BC}$-module. Indeed, as before we can define $\mathbf{j}X_2 := \mathbf{j}\mathbb{C}(\mathbf{i}) \otimes_{\mathbb{C}(\mathbf{i})} X_2$, and thus we may consider now the additive abelian group $X_1 + \mathbf{j}X_2$. Thus we are led to define the multiplication by bicomplex scalars by the formula

$$(z_1 + z_2\mathbf{j}) \cdot (x_1 + \mathbf{j}x_2) := (z_1 x_1 - z_2 x_2) + \mathbf{j}(z_1 x_2 + z_2 x_1)$$

$$= (z_1 x_1 - z_2 x_2, \mathbf{j}(z_1 x_2 + z_2 x_1))$$

but the symbols $z_1 x_1 - z_2 x_2$ and $z_1 x_2 + z_2 x_1$ are meaningless.

We could have obviously repeated the exact same procedure beginning with two $\mathbb{C}(\mathbf{j})$-linear spaces, and construcing a new $\mathbb{BC}$-module from them. This process, however, would not work if one starts with two hyperbolic modules.

Chapter 4
Norms and Inner Products on $\mathbb{BC}$-Modules

Abstract There are discussed the two types of norms on bicomplex modules: the norms with real values and those with values in non–negative hyperbolic numbers. It turns out that hyperbolic valued norms are better compatible with the structure of bicomplex modules. In particular, the bicomplex valued inner products generate in a usual way: hyperbolic, not real, valued norms. As an example, the ring of biquaternions (complex quaternions) is treated as a bicomplex module.

Keywords Bicomplex modules · Bicomplex inner products · Biquaternions · Hyperbolic valued norms

4.1 Real-Valued Norms on Bicomplex Modules

In this section we consider a norm on a bicomplex module which extends the usual properties of the Euclidean norm on $\mathbb{BC}$. Another approach that generalizes the notion of $\mathbb{D}$-valued norm on $\mathbb{BC}$ will be considered in the following section.

Definition 4.1.1 Let X be a $\mathbb{BC}$-module and let $\| \cdot \|$ be a norm on X seen as a real linear space. We say that $\| \cdot \|$ is a (real-valued) norm on the $\mathbb{BC}$-module X if for any $\mu \in \mathbb{BC}$ it is

$$\| \mu\, x \| \leq \sqrt{2}\, |\mu| \cdot \|x\| .$$

This definition extends obviously property (1.2) but it does not enjoy necessarily the other properties of the Euclidean norm. This leads us to more definitions.

Definition 4.1.2 A norm $\| \cdot \|$ on a $\mathbb{BC}$-module X is called a norm of $\mathbb{C}(\mathbf{i})$-type if

$$\| \lambda\, x \| = |\lambda| \cdot \|x\| \quad \text{for all } \lambda \in \mathbb{C}(\mathbf{i}), \text{ for all } x \in X .$$

D. Alpay et al., *Basics of Functional Analysis with Bicomplex Scalars, and Bicomplex Schur Analysis*, SpringerBriefs in Mathematics, DOI: 10.1007/978-3-319-05110-9_4, © The Author(s) 2014

Definition 4.1.3 A norm $\|\cdot\|$ on a $\mathbb{BC}$-module X is called a norm of $\mathbb{C}(\mathbf{j})$-type if

$$\|\gamma x\| = |\gamma| \cdot \|x\| \quad \text{for all } \gamma \in \mathbb{C}(\mathbf{j}), \text{ for all } x \in X.$$

Definition 4.1.4 A norm $\|\cdot\|$ on a $\mathbb{BC}$-module X is called a norm of complex type if it is simultaneously of $\mathbb{C}(\mathbf{i})$-type and of $\mathbb{C}(\mathbf{j})$-type.

Example 4.1.5 The Euclidean norm on $\mathbb{BC}$ is a norm of complex type. Other obvious examples are the usual norms on $\mathbb{BC}^n$ and on $L_p(\Omega, \mathbb{BC})$ with $p > 1$.

4.1.6. Let us take up again the situation of Sect. 3.2, that is, we are given two $\mathbb{C}(\mathbf{i})$-linear spaces X_1, X_2, and $X = \mathbf{e}X_1 + \mathbf{e}^\dagger X_2$. Now assume additionally that X_1 and X_2 are normed spaces with respective norms $\|\cdot\|_1, \|\cdot\|_2$. For any $x = \mathbf{e}x_1 + \mathbf{e}^\dagger x_2 \in X$, set

$$\|x\|_X := \frac{1}{\sqrt{2}}\sqrt{\|x_1\|_1^2 + \|x_2\|_2^2}. \tag{4.1}$$

We can show that (4.1) defines a norm on X in the sense of Definition 4.1.1. We will refer to it as the Euclidean-type norm on $X = \mathbf{e}X_1 + \mathbf{e}^\dagger X_2$. It is well-known that (4.1) defines a real norm on the real space X. But in addition we have, for any $\lambda = \lambda_1\mathbf{e} + \lambda_2\mathbf{e}^\dagger \in \mathbb{BC}, \lambda_1, \lambda_2 \in \mathbb{C}(\mathbf{i})$ and any $x = \mathbf{e}x_1 + \mathbf{e}^\dagger x_2 \in X$:

$$\|\lambda x\|_X = \|\mathbf{e}(\lambda_1 x_1) + \mathbf{e}^\dagger(\lambda_2 x_2)\|_X = \frac{1}{\sqrt{2}}\sqrt{\|\lambda_1 x_1\|_1^2 + \|\lambda_2 x_2\|_2^2}$$

$$= \frac{1}{\sqrt{2}}\sqrt{|\lambda_1|^2 \cdot \|x_1\|_1^2 + |\lambda_2|^2 \cdot \|x_2\|_2^2}$$

$$\leq \frac{1}{\sqrt{2}}\sqrt{2 \cdot \|x\|_X^2 (|\lambda_1|^2 + |\lambda_2|^2)}$$

$$= \|x\|_X \cdot \sqrt{|\lambda_1|^2 + |\lambda_2|^2} = \sqrt{2}|\lambda| \cdot \|x\|_X.$$

Since obviously for any $\lambda \in \mathbb{C}(\mathbf{i})$ we have that

$$\|\lambda x\|_X = |\lambda| \cdot \|x\|_X,$$

then (4.1) defines a norm of $\mathbb{C}(\mathbf{i})$-type. Paradoxically, (4.1) proves to be not only a $\mathbb{C}(\mathbf{i})$-norm but a $\mathbb{C}(\mathbf{j})$-norm as well. Indeed, take $\lambda = a+\mathbf{j}b = (a-\mathbf{i}b)\mathbf{e}+(a+\mathbf{i}b)\mathbf{e}^\dagger \in \mathbb{C}(\mathbf{j}) \subset \mathbb{BC}, a, b \in \mathbb{R}$, then

$$\|\lambda x\|_X = \|\mathbf{e}(a - \mathbf{i}b)x_1 + \mathbf{e}^\dagger(a + \mathbf{i}b)x_2\|_X$$

$$= \frac{1}{\sqrt{2}}\sqrt{|a - \mathbf{i}b|^2 \cdot \|x_1\|_1^2 + |a + \mathbf{i}b|^2 \cdot \|x_2\|_2^2}$$

$$= |\lambda| \cdot \|x\|_X .$$

In the same way, a pair of $\mathbb{C}(\mathbf{j})$-linear normed spaces generates a bicomplex module with a norm of $\mathbb{C}(\mathbf{j})$-type, which again is also of $\mathbb{C}(\mathbf{i})$-type.

Thus any pair of complex normed spaces generates a bicomplex module, with a norm of complex type.

4.2 $\mathbb{D}$-Valued Norm on $\mathbb{BC}$-Modules

Definition 4.2.1 Let X be a $\mathbb{BC}$-module. A map

$$\|\cdot\|_\mathbb{D} : X \longrightarrow \mathbb{D}^+$$

is said to be a hyperbolic norm on X if it satisfies the following properties:

(I) $\|x\|_\mathbb{D} = 0$ if and only if $x = 0$;

(II) $\|\mu x\|_\mathbb{D} = |\mu|_\mathbf{k} \cdot \|x\|_\mathbb{D}$ $\forall x \in X, \forall \mu \in \mathbb{BC}$;

(III) $\|x + y\|_\mathbb{D} \preccurlyeq \|x\|_\mathbb{D} + \|y\|_\mathbb{D}$ $\forall x, y \in X$.

4.2.2. If X is of the form $X = \mathbf{e} \cdot X_1 + \mathbf{e}^\dagger \cdot X_2$, with X_1 and X_2 being two arbitrary $\mathbb{C}(\mathbf{i})$-linear normed spaces with norms $\|\cdot\|_1$ and $\|\cdot\|_2$, then X can be endowed canonically with the hyperbolic norm by the formula

$$\|x\|_\mathbb{D} = \|\mathbf{e}x_1 + \mathbf{e}^\dagger x_2\|_\mathbb{D} := \|x_1\|_1 \cdot \mathbf{e} + \|x_2\|_2 \cdot \mathbf{e}^\dagger. \qquad (4.2)$$

That this is a hyperbolic norm is easily verified as follows:

$$\|\mu x\|_\mathbb{D} = \|(\mu_1 \mathbf{e} + \mu_2 \mathbf{e}^\dagger) \cdot (\mathbf{e} \cdot x_1 + \mathbf{e}^\dagger \cdot x_2)\|_\mathbb{D}$$

$$= \|\mathbf{e}(\mu_1 x_1) + \mathbf{e}^\dagger(\mu_2 x_2)\|_\mathbb{D}$$

$$= \|\mu_1 x_1\|_1 \cdot \mathbf{e} + \|\mu_2 x_2\|_2 \cdot \mathbf{e}^\dagger$$

$$= |\mu_1| \cdot \|x_1\|_1 \cdot \mathbf{e} + |\mu_2| \cdot \|x_2\|_2 \cdot \mathbf{e}^\dagger$$

$$= (|\mu_1| \cdot \mathbf{e} + |\mu_2| \cdot \mathbf{e}^\dagger) \cdot (\|x_1\|_1 \cdot \mathbf{e} + \|x_2\|_2 \cdot \mathbf{e}^\dagger)$$

$$= |\mu|_\mathbf{k} \cdot \|x\|_\mathbb{D};$$

$$\|u + v\|_{\mathbb{D}} = \|(\mathbf{e} \cdot u_1 + \mathbf{e}^\dagger \cdot u_2) + (\mathbf{e} \cdot w_1 + \mathbf{e}^\dagger \cdot w_2)\|_{\mathbb{D}}$$

$$= \|\mathbf{e}(u_1 + w_1) + \mathbf{e}^\dagger(u_2 + w_2)\|_{\mathbb{D}}$$

$$= \|u_1 + w_1\|_1 \cdot \mathbf{e} + \|u_2 + w_2\|_2 \cdot \mathbf{e}^\dagger$$

$$\preccurlyeq (\|u_1\|_1 + \|w_1\|_1) \cdot \mathbf{e} + (\|u_2\|_2 + \|w_2\|_2) \cdot \mathbf{e}^\dagger$$

$$= \|u\|_{\mathbb{D}} + \|v\|_{\mathbb{D}}.$$

We can compare (4.2) with the (real-valued) norm given by formula (4.1), and we easily see that

$$|\|x\|_{\mathbb{D}}| = |\mathbf{e} \cdot \|x_1\|_1 + \mathbf{e}^\dagger \cdot \|x_2\|_2|$$

$$= \frac{1}{\sqrt{2}} \cdot \sqrt{\|x_1\|_1^2 + \|x_2\|_2^2} \tag{4.3}$$

$$= \|x\|_X ,$$

exactly as it happens when $X = \mathbb{BC}$.

Let us now consider a $\mathbb{BC}$-module X endowed with a real-valued norm $\|\cdot\|$ of complex type. Since $X = \mathbf{e} \cdot X_{\mathbf{e}} + \mathbf{e}^\dagger \cdot X_{\mathbf{e}^\dagger}$ where $X_{\mathbf{e}}$ and $X_{\mathbf{e}^\dagger}$ are $\mathbb{C}(\mathbf{i})$- and $\mathbb{C}(\mathbf{j})$-linear spaces, the restrictions of the norm on X onto $X_{\mathbf{e}}$ and $X_{\mathbf{e}^\dagger}$ respectively:

$$\|\cdot\|_{X_{\mathbf{e}}} := \|\cdot\| \mid_{X_{\mathbf{e}}} \qquad \text{and} \qquad \|\cdot\|_{X_{\mathbf{e}^\dagger}} := \|\cdot\| \mid_{X_{\mathbf{e}^\dagger}}$$

induce norms on the complex spaces $X_{\mathbf{e}}$ and $X_{\mathbf{e}^\dagger}$ respectively. These norms, in accordance with the process outlined before, allow us to endow the bicomplex module X with a new hyperbolic norm, given by the formula

$$\|x\|_{\mathbb{D}} := \|\mathbf{e}x\|_{X_{\mathbf{e}}} \cdot \mathbf{e} + \|\mathbf{e}^\dagger x\|_{X_{\mathbf{e}^\dagger}} \cdot \mathbf{e}^\dagger. \tag{4.4}$$

Conversely, if the $\mathbb{BC}$-module X has a hyperbolic norm $\|\cdot\|_{\mathbb{D}}$, then the formula

$$\|x\|_X := |\|x\|_{\mathbb{D}}| \tag{4.5}$$

defines a norm on X. We now show that in fact (4.5) defines a usual norm in the real space X. To do so we observe that:

(a) $\|x\|_X = 0$ if and only if $\|x\|_{\mathbb{D}} = 0$ which holds if and only if $x = 0$, since $\|\cdot\|_{\mathbb{D}}$ is a hyperbolic norm;

(b) given any $\mu \in \mathbb{R}$ and $x \in X$, then

$$\|\mu x\|_X = |\|\mu x\|_{\mathbb{D}}| = \||\mu|_{\mathbf{k}} \cdot \|x\|_{\mathbb{D}}|$$

$$= \||\mu| \cdot \|x\|_{\mathbb{D}}| = |\mu| \cdot |\|x\|_{\mathbb{D}}|$$

$$= |\mu| \cdot \|x\|_X;$$

(c) for any x, $y \in X$ there holds:

$$\|x + y\|_X = |\|x + y\|_{\mathbb{D}}|$$

$$\leq \|\|x\|_{\mathbb{D}} + \|y\|_{\mathbb{D}}| \leq |\|x\|_{\mathbb{D}}| + |\|y\|_{\mathbb{D}}|$$

$$= \|x\|_X + \|y\|_X.$$

Note that we have used (1.11) in this reasoning.

Finally, we have to prove that this real norm satisfies the required inequality in Definition 4.1.1. Given $W \in \mathbb{BC}$ and $x \in X$ it is

$$\|W \cdot x\|_X = |\|W \cdot x\|_{\mathbb{D}}| = \||W|_{\mathbf{k}} \cdot \|x\|_{\mathbb{D}}|$$

$$= \||W|_{\mathbf{k}}| \cdot |\|x\|_{\mathbb{D}}|$$

$$\leq \sqrt{2} \cdot |W| \cdot \|x\|_X;$$

here we have used (1) and (2) from Remark 1.5.3.

We can summarize our arguments as follows:

Theorem 4.2.3 A $\mathbb{BC}$-module X has a real-valued norm if and only if it has a hyperbolic norm, and these norms are connected by equality (4.5).

As we already saw in the case of $\mathbb{BC}$, the two norms on a $\mathbb{BC}$-module X (the real valued norm $\|\cdot\|_X$ and the hyperbolic norm $\|\cdot\|_{\mathbb{D}}$) give rise to equivalent notions of convergence.

Definition 4.2.4 A sequence $\{x_n\}_{n \in \mathbb{N}}$ in X converges to $x_0 \in X$ with respect to the norm $\|\cdot\|_X$ if for every $\epsilon > 0$ there exists $N \in \mathbb{N}$ such that for any $n \geq N$ it is:

$$\|x_n - x_0\|_X < \epsilon.$$

Similarly for the convergence with respect to the hyperbolic norm.

Definition 4.2.5 A sequence $\{x_n\}_{n \in \mathbb{N}}$ in X converges to $x_0 \in X$ with respect to the hyperbolic norm $\|\cdot\|_{\mathbb{D}}$ if $\forall \epsilon > 0$ there exists $N \in \mathbb{N}$ such that $\forall n \geq N$ there holds:

$$\|x_n - x_0\|_{\mathbb{D}} < \epsilon.$$

From equalities (4.3), (4.4) and (4.5) it follows that a sequence $\{x_n\}_{n\in\mathbb{N}}$ converges to $x_0 \in X$ with respect to the norm $\| \cdot \|_X$ if and only if it converges to x_0 with respect to the hyperbolic norm $\| \cdot \|_{\mathbb{D}}$.

4.3 Bicomplex Modules with Inner Product

Definition 4.3.1 (Inner product). Let X be a $\mathbb{BC}$-module. A mapping

$$\langle \cdot, \cdot \rangle : X \times X \to \mathbb{BC}$$

is said to be a $\mathbb{BC}$-inner, or $\mathbb{BC}$-scalar, product on X if it satisfies the following properties:

1. $\langle x, y + z \rangle = \langle x, y \rangle + \langle x, z \rangle$ for all $x, y, z \in X$;

2. $\langle \mu x, y \rangle = \mu \langle x, y \rangle$ for all $\mu \in \mathbb{BC}$, for all $x, y \in X$;

3. $\langle x, y \rangle = \langle y, x \rangle^*$ for all $x, y \in X$;

4. $\langle x, x \rangle \in \mathbb{D}^+$, and $\langle x, x \rangle = 0$ if and only if $x = 0$.

Note that property (3) implies the following:

$$a + b\mathbf{j} = \langle x, x \rangle = \langle x, x \rangle^* = \overline{a} - \overline{b}\mathbf{j} \quad \text{with} \quad a, b \in \mathbb{C}(\mathbf{i}),$$

hence $a \in \mathbb{R}$, $b = \mathbf{i}b_1$, with $b_1 \in \mathbb{R}$, that is, $\langle x, x \rangle \in \mathbb{D}$. For reasons that will be clear later, we are strengthening this property, by requiring additionally that $\langle x, x \rangle$ will be not only in $\mathbb{D}$ but in $\mathbb{D}^+$, the set of "positive" hyperbolic numbers.

Also from Properties (2) and (3), we see that for any $\mu \in \mathbb{BC}$ and any $x, y \in X$ one has:

$$\langle x, \mu y \rangle = \mu^* \langle x, y \rangle .$$

Definition 4.3.2 A $\mathbb{BC}$-module X endowed with a bicomplex inner product $\langle \cdot, \cdot \rangle$ is said to be a $\mathbb{BC}$-inner product module.

4.3.3. We consider now, as in Sect. 3.2 and 4.1.6, two $\mathbb{C}(\mathbf{i})$-linear spaces X_1 and X_2. In addition we will assume that both are inner product spaces, with inner products $\langle \cdot, \cdot \rangle_1$ and $\langle \cdot, \cdot \rangle_2$, and corresponding norms $\| \cdot \|_1$ and $\| \cdot \|_2$. We can then show that the formula

$$\langle x, w \rangle_X = \langle \mathbf{e}x_1 + \mathbf{e}^\dagger x_2, \mathbf{e}w_1 + \mathbf{e}^\dagger w_2 \rangle_X$$

$$:= \mathbf{e}\langle x_1, w_1 \rangle_1 + \mathbf{e}^\dagger \langle x_2, w_2 \rangle_2$$

(4.6)

defines a bicomplex inner product on the bicomplex module $X = \mathbf{e}X_1 + \mathbf{e}^\dagger X_2$. We begin by proving distributivity.

1.

$$\langle x, y + w \rangle_X = \langle \mathbf{e}x_1 + \mathbf{e}^\dagger x_2, (\mathbf{e}y_1 + \mathbf{e}^\dagger y_2)$$

$$+ (\mathbf{e}w_1 + \mathbf{e}^\dagger w_2) \rangle_X$$

$$= \langle \mathbf{e}x_1 + \mathbf{e}^\dagger x_2, \mathbf{e}(y_1 + w_1) + \mathbf{e}^\dagger(y_2 + w_2) \rangle_X$$

$$= \mathbf{e}\langle x_1, y_1 + w_1 \rangle_1 + \mathbf{e}^\dagger \langle x_2, y_2 + w_2 \rangle_2$$

$$= \left(\mathbf{e}\langle x_1, y_1 \rangle_1 + \mathbf{e}^\dagger \langle x_2, y_2 \rangle_2 \right)$$

$$+ \left(\mathbf{e}\langle x_1, w_1 \rangle_1 + \mathbf{e}^\dagger \langle x_2, w_2 \rangle_2 \right)$$

$$= \langle x, y \rangle_X + \langle x, w \rangle_X .$$

2. Next, we deal with the linearity. Given $\mu = \mu_1 \mathbf{e} + \mu_2 \mathbf{e}^\dagger \in \mathbb{BC}$, then

$$\langle \mu x, y \rangle_X = \langle (\mu_1 \mathbf{e} + \mu_2 \mathbf{e}^\dagger)(\mathbf{e}x_1 + \mathbf{e}^\dagger x_2), \mathbf{e}y_1 + \mathbf{e}^\dagger y_2 \rangle_X$$

$$= \langle \mathbf{e}(\mu_1 x_1) + \mathbf{e}^\dagger(\mu_2 x_2), \mathbf{e}y_1 + \mathbf{e}^\dagger y_2 \rangle_X$$

$$= \mathbf{e}\langle \mu_1 x_1, y_1 \rangle_1 + \mathbf{e}^\dagger \langle \mu_2 x_2, y_2 \rangle_2$$

$$= \mathbf{e}\mu_1 \langle x_1, y_1 \rangle_1 + \mathbf{e}^\dagger \mu_2 \langle x_2, y_2 \rangle_2$$

$$= \left(\mathbf{e}\mu_1 + \mathbf{e}^\dagger \mu_2 \right)(\mathbf{e}\langle x_1, y_1 \rangle_1 + \mathbf{e}^\dagger \langle x_2, y_2 \rangle_2)$$

$$= \mu \langle x, y \rangle_X .$$

3. The next step is to prove what the analog of being "Hermitian for an inner product" is in our situation.

$$\langle y, x \rangle_X^* = \langle \mathbf{e}y_1 + \mathbf{e}^\dagger y_2, \mathbf{e}x_1 + \mathbf{e}^\dagger x_2 \rangle_X^*$$

$$= \left(\mathbf{e}\langle y_1, x_1 \rangle_1 + \mathbf{e}^\dagger \langle y_2, x_2 \rangle_2 \right)^*$$

$$= \mathbf{e}\overline{\langle y_1, x_1 \rangle_1} + \mathbf{e}^\dagger \overline{\langle y_2, x_2 \rangle_2}$$

$$= \mathbf{e}\langle x_1, y_1 \rangle_1 + \mathbf{e}^\dagger \langle x_2, y_2 \rangle_2 = \langle x, y \rangle_X .$$

Note that the "inner product square"

$$\langle x, x \rangle_X = \mathbf{e}\|x_1\|_1^2 + \mathbf{e}^\dagger \|x_2\|_2^2 \in \mathbb{D}^+. \tag{4.7}$$

is always a positive hyperbolic number, without a need for any additional requirements on $\langle \cdot, \cdot \rangle_1$ and $\langle \cdot, \cdot \rangle_2$. The marvelous fact here is that with the above described process, we constructed a class of bicomplex modules with inner product in such a way that property (4.7) always verifies.

4. Finally, we prove the non-degeneracy.

$$\langle x, x \rangle_X = 0 \iff \mathbf{e}\|x_1\|_1^2 + \mathbf{e}^\dagger \|x_2\|_2^2 = 0 \iff$$

$$x_1 = 0, \ x_2 = 0 \iff x = 0.$$

As already noted, the "inner product square" is a hyperbolic positive number, and this suggests the possibility of introducing a hyperbolic norm on an inner product $\mathbb{BC}$-module consistent with the bicomplex inner product. Let us show that this is, indeed, possible. Set

$$\|x\|_\mathbb{D} = \|\mathbf{e} \cdot x_1 + \mathbf{e}^\dagger \cdot x_2\|_\mathbb{D} := \langle x, x \rangle^{1/2}, \tag{4.8}$$

where $\langle x, x \rangle^{1/2}$ is the hyperbolic number given by

$$\langle x, x \rangle^{1/2} = \left(\mathbf{e}\langle x_1, x_1 \rangle_1 + \mathbf{e}^\dagger \langle x_2, x_2 \rangle_2 \right)^{1/2}$$

$$= \mathbf{e}\langle x_1, x_1 \rangle_1^{1/2} + \mathbf{e}^\dagger \langle x_2, x_2 \rangle_2^{1/2}$$

$$= \mathbf{e} \cdot \|x_1\|_1 + \mathbf{e}^\dagger \|x_2\|_2 \in \mathbb{D}^+.$$

It is easy to verify that the hyperbolic norm satisfies the properties one would expect.

(I) $\|x\|_\mathbb{D} = \|\mathbf{e}x_1 + \mathbf{e}^\dagger x_2\|_\mathbb{D} = 0$ if and only if $x_1 = 0$ and $x_2 = 0$.

(II) Given $\mu = \mu_1 \mathbf{e} + \mu_2 \mathbf{e}^\dagger \in \mathbb{BC}$, one has:

$$\|\mu x\|_\mathbb{D} = \langle \mu x, \mu x \rangle^{1/2}$$

$$= (\mu \mu^*)^{1/2} \cdot \langle x, x \rangle^{1/2}$$

$$= \left(|\mu_1| \mathbf{e} + |\mu_2| \mathbf{e}^\dagger \right) \cdot \|x\|_\mathbb{D}$$

$$= |\mu|_\mathbf{k} \cdot \|x\|_\mathbb{D}.$$

(III) For any $x, y \in X$, it follows that:

$$\|x + y\|_{\mathbb{D}} = \langle x + y, x + y \rangle^{1/2}$$

$$= \left(\langle x_1 + y_1, x_1 + y_1 \rangle_1 \cdot \mathbf{e} + \langle x_2 + y_2, x_2 + y_2 \rangle_2 \cdot \mathbf{e}^\dagger \right)^{1/2}$$

$$= \left(\|x_1 + y_1\|_1^2 \cdot \mathbf{e} + \|x_2 + y_2\|_2^2 \cdot \mathbf{e}^\dagger \right)^{1/2}$$

$$\preccurlyeq \left((\|x_1\|_1 + \|y_1\|_1)^2 \cdot \mathbf{e} + (\|x_2\|_2 + \|y_2\|_2)^2 \cdot \mathbf{e}^\dagger \right)^{1/2}$$

$$= \|x\|_{\mathbb{D}} + \|y\|_{\mathbb{D}}.$$

4.3.4. On the other hand, it is also possible to endow the bicomplex module $X = \mathbf{e} X_1 + \mathbf{e}^\dagger X_2$ with a real-valued norm naturally related to the bicomplex inner product discussed above. For this purpose, we simply set

$$\|x\|_X^2 := \frac{1}{2} \left(\langle x_1, x_1 \rangle_1 + \langle x_2, x_2 \rangle_2 \right)$$

$$= \frac{1}{2} \left(\|x_1\|_1^2 + \|x_2\|_2^2 \right),$$

(4.9)

and we compare this definition with formulas (4.1) and (4.7); we immediately see that (4.9) defines a Euclidean-type norm on X which is related to the norms on X_1 and X_2 exactly in the same way as the Euclidean norm on $\mathbb{BC}$ is related to the modules of the coefficients of its idempotent representation. The relation between (4.9) and (4.8) is as one would expect:

$$\left| \|x\|_{\mathbb{D}} \right| = \left| \mathbf{e} \cdot \|x_1\|_1 + \mathbf{e}^\dagger \cdot \|x_2\|_2 \right|$$

$$= \frac{1}{\sqrt{2}} \cdot \sqrt{\|x_1\|_1^2 + \|x_2\|_2^2} = \|x\|_X.$$

Definition 4.3.5 A $\mathbb{BC}$-inner product module X is said to be a bicomplex Hilbert module if it is complete with respect to the metric induced by its Euclidean-type norm generated by the inner product; this is equivalent to say that X is complete with respect to the hyperbolic norm generated by the inner product square.

Remark 4.3.6 We can restate this definition by saying that a bicomplex Hilbert module is a 4-tuple $(X, \| \cdot \|_X, \| \cdot \|_{\mathbb{D}}, \langle \cdot, \cdot \rangle)$ where any Cauchy sequence is convergent.

It follows from (4.9) that if $X = \mathbf{e} X_1 + \mathbf{e}^\dagger X_2$, where X_1, X_2 are $\mathbb{C}(\mathbf{i})$-linear inner product spaces, then X is a bicomplex Hilbert space if and only if X_1 and X_2 are complex Hilbert spaces.

Example 4.3.7 A bicomplex inner product in $\mathbb{BC}$ can be given by the formula:

$$\langle Z, W \rangle := Z W^*. \tag{4.10}$$

Properties (1), (2), and (3) of Definition 4.3.1 are obvious. For the fourth property recall that if $Z = z_1 + \mathbf{j} z_2 = \beta_1 \cdot \mathbf{e} + \beta_2 \cdot \mathbf{e}^\dagger$, then

$$\langle Z, Z \rangle = Z Z^*$$

$$= (\beta_1 \cdot \mathbf{e} + \beta_2 \cdot \mathbf{e}^\dagger) \cdot (\overline{\beta}_1 \cdot \mathbf{e} + \overline{\beta}_2 \cdot \mathbf{e}^\dagger)$$

$$= |\beta_1| \cdot \mathbf{e} + |\beta_2| \cdot \mathbf{e}^\dagger \tag{4.11}$$

$$= \left(|z_1|^2 + |z_2|^2 \right) - 2 Im(z_1 \overline{z}_2) \, \mathbf{i} \mathbf{j} \in \mathbb{D}^+.$$

Thus $\langle Z, Z \rangle = 0$ if and only if $z_1 = 0 = z_2$ or equivalently $Z = 0$. Hence the general definition of the inner product is simply a generalization of this definition for the bicomplex-valued inner product in $\mathbb{BC}$. Note in particular that non-degeneracy would be lost if we had used any of the two other conjugations on $\mathbb{BC}$. Moreover, the inner product square of any element is a hyperbolic number in $\mathbb{D}^+$. This is the reason that, in [1], it is required that the inner product square of any element (in an arbitrary bicomplex inner product module) had to be a positive hyperbolic number. Note that in this example the real part of the hyperbolic number (4.11) is already a good norm.

Let us now compare these two approaches with Sect. 4.3.3. The idempotent decomposition of $\mathbb{BC}$:

$$\mathbb{BC} = \mathbb{C}(\mathbf{i})\mathbf{e} + \mathbb{C}(\mathbf{i})\mathbf{e}^\dagger,$$

can be reformulated, by saying that $X_1 = \mathbb{C}(\mathbf{i})\mathbf{e}$, $X_2 = \mathbb{C}(\mathbf{i})\mathbf{e}^\dagger$ and with the $\mathbb{C}(\mathbf{i})$-valued inner products

$$\langle \beta_1 \mathbf{e}, \beta_1' \mathbf{e} \rangle_1 := \beta_1 \overline{\beta_1'} \quad \forall \beta_1 \mathbf{e}, \, \beta_1' \mathbf{e} \in X_1, \, \beta_1, \, \beta_1' \in \mathbb{C}(\mathbf{i}),$$

$$\langle \beta_2 \mathbf{e}^\dagger, \beta_2' \mathbf{e}^\dagger \rangle_2 := \beta_2 \overline{\beta_2'} \quad \forall \beta_2 \mathbf{e}^\dagger, \, \beta_2' \mathbf{e}^\dagger \in X_2, \, \beta_2, \, \beta_2' \in \mathbb{C}(\mathbf{i}).$$

Now, in accordance with (4.6), if we take $Z = \beta_1 \mathbf{e} + \beta_2 \mathbf{e}^\dagger$, and $W = \beta_1' \mathbf{e} + \beta_2' \mathbf{e}^\dagger$, we obtain that

$$\langle Z, W \rangle_X := \langle \beta_1 \mathbf{e}, \beta_1' \mathbf{e} \rangle_1 \mathbf{e} + \langle \beta_2 \mathbf{e}^\dagger, \beta_2' \mathbf{e}^\dagger \rangle_2 \mathbf{e}^\dagger,$$

and therefore

$$\langle Z, W \rangle_X = \beta_1 \overline{\beta_1'} \mathbf{e} + \beta_2 \overline{\beta_2'} \mathbf{e}^\dagger = \left(\beta_1 \mathbf{e} + \beta_2 \mathbf{e}^\dagger \right) \left(\overline{\beta_1'} \mathbf{e} + \overline{\beta_2'} \mathbf{e}^\dagger \right)$$

$$= Z W^* = \langle Z, W \rangle.$$

This means that the inner product (4.10) on $\mathbb{BC}$ coincides with the one generated by its idempotent decomposition. In addition, the Euclidean norm on $\mathbb{BC}$ coincides with the norm on $\mathbb{BC}$ given by formula (4.9). In other words the Euclidean norm of bicomplex numbers is the Euclidean-type norm on the bicomplex inner product module $\mathbb{BC}$.

Finally, if one takes, in (4.11), Z in the idempotent form $Z = \beta_1 \mathbf{e} + \beta_2 \mathbf{e}^\dagger$ then the inner product square is

$$\langle Z, Z \rangle_X = Z \cdot Z^* = |Z|_{\mathbf{k}}^2 = |\beta_1|^2 \cdot \mathbf{e} + |\beta_2|^2 \cdot \mathbf{e}^\dagger$$

giving us a hyperbolic norm on the $\mathbb{BC}$-module $\mathbb{BC}$.

4.4 Inner Products and Cartesian Decompositions

4.4.1. We will now assume that a bicomplex inner product module X has a bar-involution, and we denote by $X_{1,bar}$ the $\mathbb{C}(\mathbf{j})$-linear space of its $\mathbb{C}(\mathbf{j})$-elements. Thus $X = X_{1,bar} + \mathbf{i} X_{1,bar}$ and $X_{\mathbb{C}(\mathbf{j})} = X_{1,bar} \oplus_{\mathbb{C}(\mathbf{j})} \mathbf{i} X_{1,bar}$, when we restrict the scalar multiplication to $\mathbb{C}(\mathbf{j})$. Nevertheless, the inner product takes values in $\mathbb{BC}$. If we now define an inner product

$$\langle \cdot, \cdot \rangle_{bar} : X_{\mathbb{C}(\mathbf{j})} \times X_{\mathbb{C}(\mathbf{j})} \to \mathbb{C}(\mathbf{j})$$

by

$$\langle x, y \rangle_{bar} := \Pi_{1,\mathbf{j}} (\langle x, y \rangle) \in \mathbb{C}(\mathbf{j}),$$

we see that $\langle \cdot, \cdot \rangle_{bar}$ is an inner product on the $\mathbb{C}(\mathbf{j})$-linear space $X_{\mathbb{C}(\mathbf{j})}$, and that it induces an inner product on the $\mathbb{C}(\mathbf{j})$-linear space $X_{1,bar}$ as follows:

$$\langle x_1, y_1 \rangle_{\mathbb{C}(\mathbf{j})} := \langle x_1, y_1 \rangle_{bar} \quad \forall x_1, y_1 \in X_{1,bar}.$$

The relation between both inner products, the bicomplex one and the $\mathbb{C}(\mathbf{j})$-valued one, is established by the following formula: if $x = x_1 + \mathbf{i} x_2$, $y = y_1 + \mathbf{i} y_2$ are in X, then

$$\langle x, y \rangle = \langle x_1 + \mathbf{i} x_2, y_1 + \mathbf{i} y_2 \rangle$$

$$= \langle x_1, y_1 \rangle_{\mathbb{C}(\mathbf{j})} + \langle x_2, y_2 \rangle_{\mathbb{C}(\mathbf{j})} \tag{4.12}$$

$$+ \mathbf{i} \left(\langle x_2, y_1 \rangle_{\mathbb{C}(\mathbf{j})} - \langle x_1, y_2 \rangle_{\mathbb{C}(\mathbf{j})} \right).$$

This formula provides even more information. Indeed, if X is just a bicomplex module, without any bicomplex inner product, and if $X_{1,bar}$ has a $\mathbb{C}(\mathbf{j})$-valued inner product, then formula (4.12) tells us how to endow X with a bicomplex product which extends the one on $X_{1,bar}$.

4.4.2. We can treat in exactly the same way the case in which X admits a †-involution. In this case $X = X_{1,†} + \mathbf{j}\,X_{1,†}$, and $X_{\mathbb{C}(\mathbf{i})} = X_{1,†} \oplus_{\mathbb{C}(\mathbf{i})} \mathbf{j}\,X_{1,†}$. The mapping

$$\langle \cdot, \cdot \rangle_† : X_{\mathbb{C}(\mathbf{i})} \times X_{\mathbb{C}(\mathbf{i})} \to \mathbb{C}(\mathbf{i})$$

given by

$$\langle x, y \rangle_† := \Pi_{1,\mathbf{i}}\left(\langle x, y \rangle \right) \in \mathbb{C}(\mathbf{i})$$

is an inner product on the $\mathbb{C}(\mathbf{i})$-linear space $X_{\mathbb{C}(\mathbf{i})}$. The induced inner product on the $\mathbb{C}(\mathbf{i})$-linear space $X_{1,†}$ for any $s_1, t_1 \in X_{1,†}$ is:

$$\langle s_1, t_1 \rangle_{\mathbb{C}(\mathbf{i})} := \langle s_1, t_1 \rangle_† .$$

Now, if $x = s_1 + \mathbf{j}\,s_2$, and $y = t_1 + \mathbf{j}\,t_2$ are in X then

$$
\begin{aligned}
\langle x, y \rangle &= \langle s_1 + \mathbf{j}\,s_2, t_1 + \mathbf{j}\,t_2 \rangle \\
&= \langle s_1, t_1 \rangle_{\mathbb{C}(\mathbf{i})} + \langle s_2, t_2 \rangle_{\mathbb{C}(\mathbf{i})} + \mathbf{j}\left(\langle s_2, t_1 \rangle_{\mathbb{C}(\mathbf{i})} - \langle s_1, t_2 \rangle_{\mathbb{C}(\mathbf{i})} \right).
\end{aligned}
\tag{4.13}
$$

Again, formula (4.13) provides an extension of a $\mathbb{C}(\mathbf{i})$-valued inner product defined on $X_{1,†}$ up to a bicomplex inner product on X.

Example 4.4.3 We continue now the analysis iniciated in Example 4.3.7. The space $X = \mathbb{BC}$ has both involutions, so that $X_{1,bar} = \mathbb{C}(\mathbf{j})$ and $X_{1,†} = \mathbb{C}(\mathbf{i})$. Given $Z, W \in \mathbb{BC}$, with $Z = z_1 + \mathbf{j}\,z_2 = s_1 + \mathbf{i}\,s_2$ and $W = w_1 + \mathbf{j}\,w_2 = t_1 + \mathbf{i}\,t_2$, z_1, z_2, $w_1, w_2 \in \mathbb{C}(\mathbf{i})$, $s_1, s_2, t_1, t_2 \in \mathbb{C}(\mathbf{j})$, then

$$
\begin{aligned}
\langle Z, W \rangle &= ZW^* = (z_1 + \mathbf{j}\,z_2) \cdot (w_1 + \mathbf{j}\,w_2)^* \\[4pt]
&= (z_1 + \mathbf{j}\,z_2) \cdot (\overline{w}_1 - \mathbf{j}\,\overline{w}_2) \\[4pt]
&= (z_1\,\overline{w}_1 + z_2\,\overline{w}_2) + \mathbf{j}\,(z_2\,\overline{w}_1 - z_1\,\overline{w}_2) \\[4pt]
&= (s_1 + \mathbf{i}\,s_2) \cdot (t_1 + \mathbf{i}\,t_2)^* \\[4pt]
&= (s_1 + \mathbf{i}\,s_2) \cdot \left(t_1^* - \mathbf{i}\,t_2^*\right) = \left(s_1\,t_1^* + s_2\,t_2^*\right) + \mathbf{i}\left(s_2\,t_1^* - s_1\,t_2^*\right),
\end{aligned}
$$

which implies that on the $\mathbb{C}(\mathbf{j})$-linear space $\mathbb{BC}_{\mathbb{C}(\mathbf{j})}$ the $\mathbb{C}(\mathbf{j})$-valued inner product $\langle \cdot, \cdot \rangle_{bar}$ is given by

$$\langle Z, W \rangle_{bar} = s_1\,t_1^* + s_2 t_2^* \in \mathbb{C}(\mathbf{j})$$

and on the $\mathbb{C}(\mathbf{i})$-linear space $\mathbb{BC}_{\mathbb{C}(\mathbf{i})}$ the $\mathbb{C}(\mathbf{i})$-valued inner product $\langle \cdot, \cdot \rangle_†$ is

$$\langle Z, W \rangle_† = z_1\,\overline{w}_1 + z_2\,\overline{w}_2 \in \mathbb{C}(\mathbf{i}) ;$$

finally it is

$$\langle s_1, t_1 \rangle_{\mathbb{C}(\mathbf{j})} := s_1\, t_1^*, \qquad \langle z_1, w_1 \rangle_{\mathbb{C}(\mathbf{i})} := z_1\, \overline{w}_1 \, ,$$

which are the usual inner products on $\mathbb{C}(\mathbf{j})$ and $\mathbb{C}(\mathbf{i})$.

4.5 Inner Products and Idempotent Decompositions

4.5.1. Consider a bicomplex module X with an inner product $\langle \cdot, \cdot \rangle$. For any $x, y \in X$, the inner product $\langle x, y \rangle$ is given by

$$\langle x, y \rangle = \beta_1(x, y)\, \mathbf{e} + \beta_2(x, y)\, \mathbf{e}^\dagger$$
$$= \gamma_1(x, y)\, \mathbf{e} + \gamma_2(x, y)\, \mathbf{e}^\dagger \, , \tag{4.14}$$

where $\beta_\ell(x, y) = \pi_{\ell,\mathbf{i}}(\langle x, y \rangle) \in \mathbb{C}(\mathbf{i})$, $\gamma_\ell(x, y) = \pi_{\ell,\mathbf{j}}(\langle x, y \rangle) \in \mathbb{C}(\mathbf{j})$, for $\ell = 1, 2$. Define

$$\mathcal{B}_1, \mathcal{B}_2 : X \times X \to \mathbb{C}(\mathbf{i}) \qquad \text{and} \qquad \mathcal{F}_1, \mathcal{F}_2 : X \times X \to \mathbb{C}(\mathbf{j})$$

by

$$\mathcal{B}_\ell(x, y) := \pi_{\ell,\mathbf{i}}(\langle x, y \rangle) = \beta_\ell(x, y) \in \mathbb{C}(\mathbf{i}) \tag{4.15}$$

and

$$\mathcal{F}_\ell(x, y) := \pi_{\ell,\mathbf{j}}(\langle x, y \rangle) = \gamma_\ell(x, y) \in \mathbb{C}(\mathbf{j}) \, . \tag{4.16}$$

We will prove that the restrictions

$$\langle \cdot, \cdot \rangle_{X_{\mathbf{e},\mathbf{i}}} := \mathcal{B}_1\, |_{X_\mathbf{e} \times X_\mathbf{e}} : X_\mathbf{e} \times X_\mathbf{e} \to \mathbb{C}(\mathbf{i}) \, ; \tag{4.17}$$

$$\langle \cdot, \cdot \rangle_{X_{\mathbf{e},\mathbf{j}}} := \mathcal{F}_1\, |_{X_\mathbf{e} \times X_\mathbf{e}} : X_\mathbf{e} \times X_\mathbf{e} \to \mathbb{C}(\mathbf{j}) \, ; \tag{4.18}$$

$$\langle \cdot, \cdot \rangle_{X_{\mathbf{e}^\dagger,\mathbf{i}}} := \mathcal{B}_2\, |_{X_{\mathbf{e}^\dagger} \times X_{\mathbf{e}^\dagger}} : X_{\mathbf{e}^\dagger} \times X_{\mathbf{e}^\dagger} \to \mathbb{C}(\mathbf{i}) \, ; \tag{4.19}$$

$$\langle \cdot, \cdot \rangle_{X_{\mathbf{e}^\dagger,\mathbf{j}}} := \mathcal{F}_2\, |_{X_{\mathbf{e}^\dagger} \times X_{\mathbf{e}^\dagger}} : X_{\mathbf{e}^\dagger} \times X_{\mathbf{e}^\dagger} \to \mathbb{C}(\mathbf{j}) \, , \tag{4.20}$$

are the usual inner products on the $\mathbb{C}(\mathbf{i})$- and $\mathbb{C}(\mathbf{j})$-linear spaces $X_\mathbf{e}$ and $X_{\mathbf{e}^\dagger}$ respectively. Indeed, for any $x, y, z \in X_\mathbf{e}$ and $\lambda \in \mathbb{C}(\mathbf{i})$, we have:

(a)

$$\langle x, y + z \rangle_{X_{\mathbf{e},\mathbf{i}}} = \pi_{1,\mathbf{i}}(\langle x, y + z \rangle) = \pi_{1,\mathbf{i}}(\langle x, y \rangle + \langle x, z \rangle)$$

$$= \pi_{1,\mathbf{i}}(\langle x, y \rangle) + \pi_{1,\mathbf{i}}(\langle x, z \rangle)$$

$$= \langle x, y \rangle_{X_{\mathbf{e},\mathbf{i}}} + \langle x, z \rangle_{X_{\mathbf{e},\mathbf{i}}} \, .$$

(b) Note that given $\mathbf{e}\, x, \mathbf{e}\, y \in X_\mathbf{e}$ then

$$\langle \mathbf{e}\, x, \mathbf{e}\, y \rangle = \mathbf{e}\, \mathbf{e}^* \langle x, y \rangle = \mathbf{e}\, \langle x, y \rangle \, ,$$

which implies

$$\langle \mathbf{e}\,x, \mathbf{e}\,y \rangle_{X_{\mathbf{e},i}} = \pi_{1,i}(\mathbf{e}\,\langle x, y \rangle) = \pi_{1,i}(\langle x, y \rangle) = \beta_1(x, y). \tag{4.21}$$

Similarly, we have that

$$\langle \mathbf{e}^\dagger x, \mathbf{e}^\dagger y \rangle_{X_{\mathbf{e}^\dagger,i}} = \pi_{2,i}(\langle x, y \rangle) = \beta_2(x, y), \tag{4.22}$$

and

$$\langle \mathbf{e}\,x, \mathbf{e}\,y \rangle_{X_{\mathbf{e},j}} = \pi_{1,j}(\langle x, y \rangle), \quad \langle \mathbf{e}^\dagger x, \mathbf{e}^\dagger y \rangle_{X_{\mathbf{e}^\dagger,j}} = \pi_{2,j}(\langle x, y \rangle).$$

Now, given $\lambda \in \mathbb{C}(\mathbf{i})$ and $\mathbf{e}\,x$, $\mathbf{e}\,y \in X_{\mathbf{e}}$ then

$$\langle \lambda\,(\mathbf{e}\,x), \mathbf{e}\,y \rangle_{X_{\mathbf{e},i}} = \pi_{1,i}(\langle \lambda\,\mathbf{e}\,x, \mathbf{e}\,y \rangle)$$

$$= \pi_{1,i}(\lambda\,\langle \mathbf{e}\,x, \mathbf{e}\,y \rangle) = \lambda\,\pi_{1,i}(\langle \mathbf{e}\,x, \mathbf{e}\,y \rangle)$$

$$= \lambda\,\langle \mathbf{e}\,x, \mathbf{e}\,y \rangle_{X_{\mathbf{e},i}}.$$

(c) One has:

$$\langle y, x \rangle_{X_{\mathbf{e},i}} = \pi_{1,i}\,(\langle y, x \rangle) = \pi_{1,i}\,(\langle x, y \rangle^*);$$

if we write

$$\langle x, y \rangle = \beta_1(x, y)\,\mathbf{e} + \beta_2(x, y)\,\mathbf{e}^\dagger,$$

with $\beta_1(x, y)$, $\beta_2(x, y) \in \mathbb{C}(\mathbf{i})$, then

$$\langle x, y \rangle^* = \overline{\beta_1(x, y)}\,\mathbf{e} + \overline{\beta_2(x, y)}\,\mathbf{e}^\dagger,$$

and

$$\pi_{1,i}(\langle x, y \rangle^*) = \langle x, y \rangle^*_{X_{\mathbf{e},i}} = \overline{\beta_1\,(x, y)}$$

$$= \overline{\pi_{1,i}\,(\langle x, y \rangle)} = \overline{\langle x, y \rangle}_{X_{\mathbf{e},i}}.$$

(d) First note that if $x, y \in X_{\mathbf{e}}$, then $x = \mathbf{e}\,x$, $y = \mathbf{e}\,y$, and thus

$$\mathbb{BC} \ni \beta_1(x, y)\,\mathbf{e} + \beta_2(x, y)\,\mathbf{e}^\dagger = \langle x, y \rangle = \langle \mathbf{e}\,x, \mathbf{e}\,y \rangle$$

$$= \mathbf{e}\,\langle x, y \rangle = \mathbf{e}\,\left(\beta_1(x, y)\mathbf{e} + \beta_2(x, y)\mathbf{e}^\dagger \right)$$

$$= \beta_1(x, y)\,\mathbf{e}, \tag{4.23}$$

hence $\beta_2(x, y) = 0$. In particular, taking $y = x \in X_{\mathbf{e}}$ one has $\langle x, x \rangle = \beta_1(x, x)\,\mathbf{e}$. Recall that $\langle x, x \rangle = 0$ if and only if $x = 0$, that is, $x = 0$ if and only if $\beta_1(x, x) = 0$. On the other hand

$$\beta_1(x, x) = \pi_{1,\mathbf{i}}\left(\langle x, x\rangle\right) = \langle x, x\rangle_{X_{\mathbf{e},\mathbf{i}}}$$

and thus the last property of the complex inner product is proved.

Equation (4.23) gives one more reason to call the $\mathbb{C}(\mathbf{i})$-elements the elements of $X_{\mathbf{e}}$.

Quite analogously we can show that formulas (4.18), (4.19), (4.20) define (complex) inner products on the corresponding linear spaces. Thus we have just proved the following:

Theorem 4.5.2 Let X be a $\mathbb{BC}$-inner product space. Then:

(a) $\left(X_{\mathbf{e}}, \langle \cdot, \cdot\rangle_{X_{\mathbf{e},\mathbf{i}}}\right)$ and $\left(X_{\mathbf{e}^\dagger}, \langle \cdot, \cdot\rangle_{X_{\mathbf{e}^\dagger,\mathbf{i}}}\right)$ are $\mathbb{C}(\mathbf{i})$-inner product spaces.

(b) $\left(X_{\mathbf{e}}, \langle \cdot, \cdot\rangle_{X_{\mathbf{e},\mathbf{j}}}\right)$ and $\left(X_{\mathbf{e}^\dagger}, \langle \cdot, \cdot\rangle_{X_{\mathbf{e}^\dagger,\mathbf{j}}}\right)$ are $\mathbb{C}(\mathbf{j})$-inner product spaces.

4.5.3. We will now show that the results from Sect. 4.5 are fully consistent with those in Sects. 4.3.3 and 4.3.4. Indeed, since

$$X = \mathbf{e}X + \mathbf{e}^\dagger X = X_{\mathbf{e}} + X_{\mathbf{e}^\dagger} = \mathbf{e}X_{\mathbf{e}} + \mathbf{e}^\dagger X_{\mathbf{e}^\dagger} =: \mathbf{e}X_1 + \mathbf{e}^\dagger X_2,$$

with $X_1 = \mathbf{e}X = X_{\mathbf{e}}$, $X_2 = \mathbf{e}^\dagger X = X_{\mathbf{e}^\dagger}$, we can consider the bicomplex module X as generated by the $\mathbb{C}(\mathbf{i})$-linear spaces X_1 and X_2. Moreover for $x, y \in X$ one has:

$$x = \mathbf{e}x + \mathbf{e}^\dagger x = \mathbf{e}(\mathbf{e}x) + \mathbf{e}^\dagger(\mathbf{e}^\dagger x) =: \mathbf{e}x_1 + \mathbf{e}^\dagger x_2 \in \mathbf{e}X_1 + \mathbf{e}^\dagger X_2,$$

$$y = \mathbf{e}y + \mathbf{e}^\dagger y = \mathbf{e}(\mathbf{e}y) + \mathbf{e}^\dagger(\mathbf{e}^\dagger y) =: \mathbf{e}y_1 + \mathbf{e}^\dagger y_2 \in \mathbf{e}X_1 + \mathbf{e}^\dagger X_2$$

where $x_1, y_1 \in X_1$ and $x_2, y_2 \in X_2$.

By Theorem 4.5.2, we have that $(X_1, \langle \cdot, \cdot\rangle_1)$ and $(X_2, \langle \cdot, \cdot\rangle_2)$, with $\langle \cdot, \cdot\rangle_1 := \langle \cdot, \cdot\rangle_{X_{\mathbf{e},\mathbf{i}}}$ and $\langle \cdot, \cdot\rangle_2 := \langle \cdot, \cdot\rangle_{X_{\mathbf{e}^\dagger,\mathbf{i}}}$, are $\mathbb{C}(\mathbf{i})$-inner product spaces. Thus, the results in Sect. 4.3.3 imply that X can be endowed with the bicomplex inner product

$$\langle x, y\rangle_X := \mathbf{e}\langle x_1, y_1\rangle_1 + \mathbf{e}^\dagger\langle x_2, y_2\rangle_2$$

$$= \mathbf{e}\langle \mathbf{e}x, \mathbf{e}y\rangle_{X_{\mathbf{e},\mathbf{i}}} + \mathbf{e}^\dagger\langle \mathbf{e}^\dagger x, \mathbf{e}^\dagger y\rangle_{X_{\mathbf{e}^\dagger,\mathbf{i}}}.$$

It turns out that
$$\langle x, y\rangle_X = \langle x, y\rangle.$$

Indeed, using (4.21) and (4.22)

$$\langle x, y\rangle_X = \mathbf{e}\beta_1(x, y) + \mathbf{e}^\dagger\beta_2(x, y) = \langle x, y\rangle.$$

We summarize this analysis in the next theorem.

Theorem 4.5.4 Given a bicomplex module X, then

(1) X has a bicomplex inner product $\langle \cdot, \cdot \rangle$ if and only if both $X_{\mathbf{e},\mathbf{i}}$ and $X_{\mathbf{e}^\dagger,\mathbf{i}}$ have complex inner products $\langle \cdot, \cdot \rangle_{X_{\mathbf{e},\mathbf{i}}}$ and $\langle \cdot, \cdot \rangle_{X_{\mathbf{e}^\dagger,\mathbf{i}}}$; if this holds then the complex inner products are the components of the idempotent representation of the bicomplex inner product, i.e.,

$$\langle x, y \rangle = \mathbf{e}\langle \mathbf{e}x, \mathbf{e}y \rangle_{X_{\mathbf{e},\mathbf{i}}} + \mathbf{e}^\dagger \langle \mathbf{e}^\dagger x, \mathbf{e}^\dagger y \rangle_{X_{\mathbf{e}^\dagger,\mathbf{i}}}. \tag{4.24}$$

(2) If X has a bicomplex inner product then it has a real-valued Euclidean-type norm and a hyperbolic norm which are related to the norms on $X_{\mathbf{e},\mathbf{i}}$ and $X_{\mathbf{e}^\dagger,\mathbf{i}}$ by the formulas

$$\|x\|_X := \frac{1}{\sqrt{2}} \sqrt{ \|\mathbf{e}x\|_{X_{\mathbf{e},\mathbf{i}}}^2 + \|\mathbf{e}^\dagger x\|_{X_{\mathbf{e}^\dagger,\mathbf{i}}}^2 }. \tag{4.25}$$

and

$$\|x\|_{\mathbb{D}} := \|\mathbf{e}x\|_{X_{\mathbf{e},\mathbf{i}}} \cdot \mathbf{e} + \|\mathbf{e}^\dagger x\|_{X_{\mathbf{e}^\dagger,\mathbf{i}}} \cdot \mathbf{e}^\dagger. \tag{4.26}$$

(3) $(X, \langle \cdot, \cdot \rangle)$ is a bicomplex Hilbert space if and only if $\left(X_{\mathbf{e},\mathbf{i}}, \langle \cdot, \cdot \rangle_{X_{\mathbf{e},\mathbf{i}}} \right)$ and $\left(X_{\mathbf{e}^\dagger,\mathbf{i}}, \langle \cdot, \cdot \rangle_{X_{\mathbf{e}^\dagger,\mathbf{i}}} \right)$ are complex Hilbert spaces.

The previous theorem deals with arbitrary bicomplex modules on which we have made no additional assumptions; the results explain the relations between the inner products, which can be defined. In Sect. 4.4, on the other hand, we developed the theory of bicomplex inner products in the case in which a bicomplex module X has a †-involution or a bar-involution. We will now show how these situations interact.

Theorem 4.5.5 Let X be a $\mathbb{BC}$-module with a †-involution. Then the following three statements are equivalent:

(1) X has a bicomplex inner product.

(2) Each of the $\mathbb{C}(\mathbf{i})$-linear spaces $X_{\mathbf{e}}$ and $X_{\mathbf{e}^\dagger}$ has a $\mathbb{C}(\mathbf{i})$-valued inner product, $\langle \cdot, \cdot \rangle_{X_{\mathbf{e},\mathbf{i}}}$ and $\langle \cdot, \cdot \rangle_{X_{\mathbf{e}^\dagger,\mathbf{i}}}$ respectively.

(3) The $\mathbb{C}(\mathbf{i})$-linear space $X_{1,\dagger}$ has a $\mathbb{C}(\mathbf{i})$-valued inner product $\langle \cdot, \cdot \rangle_{X_{1,\dagger}}$.

Proof:
We already proved that (1) is equivalent to (2) and that (1) implies (3).
Assume now that $X_{1,\dagger}$ has a $\mathbb{C}(\mathbf{i})$-valued inner product $\langle \cdot, \cdot \rangle_{\mathbb{C}(\mathbf{i})}$. Take $x = s_1 + \mathbf{j}s_2$ and $y = t_1 + \mathbf{j}t_2$ in X. By (4.13) we have that formula

$$\langle x, y \rangle := \left(\langle s_1 + \mathbf{i}s_2, t_1 \rangle_{\mathbb{C}(\mathbf{i})} + \langle s_2 - \mathbf{i}s_1, t_2 \rangle_{\mathbb{C}(\mathbf{i})} \right) \mathbf{e}$$

$$+ \left(\langle s_1 - \mathbf{i}s_2, t_1 \rangle_{\mathbb{C}(\mathbf{i})} + \langle s_2 + \mathbf{i}s_1, t_2 \rangle_{\mathbb{C}(\mathbf{i})} \right) \mathbf{e}^\dagger$$

gives a bicomplex inner product on X; thus each of its complex components defines an inner product on the corresponding idempotent components of X:

$$\langle \mathbf{e}x, \mathbf{e}y \rangle_{X_{\mathbf{e},\mathbf{i}}} := \langle s_1 + \mathbf{i}\, s_2, t_1 \rangle_{\mathbb{C}(\mathbf{i})} + \langle s_2 - \mathbf{i}\, s_1, t_2 \rangle_{\mathbb{C}(\mathbf{i})},$$

$$\langle \mathbf{e}^{\dagger}x, \mathbf{e}^{\dagger}y \rangle_{X_{\mathbf{e}^{\dagger},\mathbf{i}}} := \langle s_1 - \mathbf{i}\, s_2, t_1 \rangle_{\mathbb{C}(\mathbf{i})} + \langle s_2 + \mathbf{i}\, s_1, t_2 \rangle_{\mathbb{C}(\mathbf{i})};$$

this shows that (3) implies (1) and (2).   □

It is obvious that when X has a bar-involution then conditions (2) and (3) in the above theorem will require $\mathbb{C}(\mathbf{j})$-linearity instead of $\mathbb{C}(\mathbf{i})$-linearity.

4.6 Complex Inner Products on X Induced by Idempotent Decompositions

It is possible to give a different approach to the notion of bicomplex Hilbert space, by involving only complex structures. We briefly study this approach below.

We can consider X as $X_{\mathbb{C}(\mathbf{i})}$ and $X_{\mathbb{C}(\mathbf{j})}$; from this point of view, X is endowed with a specific inner product if this is true for both $X_{\mathbf{e},\mathbf{i}}$ and $X_{\mathbf{e}^{\dagger},\mathbf{i}}$, as well as both $X_{\mathbf{e},\mathbf{j}}$ and $X_{\mathbf{e}^{\dagger},\mathbf{j}}$, where the subindexes $\mathbf{i}$ and $\mathbf{j}$ mean $\mathbb{C}(\mathbf{i})$ or $\mathbb{C}(\mathbf{j})$ linearities. Indeed, since

$$X_{\mathbb{C}(\mathbf{i})} = X_{\mathbf{e},\mathbf{i}} \oplus_{\mathbb{C}(\mathbf{i})} X_{\mathbf{e}^{\dagger},\mathbf{i}}$$

and

$$X_{\mathbb{C}(\mathbf{j})} = X_{\mathbf{e},\mathbf{j}} \oplus_{\mathbb{C}(\mathbf{j})} X_{\mathbf{e}^{\dagger},\mathbf{j}},$$

we can endow $X_{\mathbb{C}(\mathbf{i})}$ and $X_{\mathbb{C}(\mathbf{j})}$ with the following inner products respectively: given $x, y \in X_{\mathbb{C}(\mathbf{i})}$ or in $X_{\mathbb{C}(\mathbf{j})}$ set

$$\langle x, y \rangle_{\mathbb{C}(\mathbf{i})} := \langle P(x) + Q(x), P(y) + Q(y) \rangle_{\mathbb{C}(\mathbf{i})}$$
$$:= \langle P(x), P(y) \rangle_{X_{\mathbf{e},\mathbf{i}}} + \langle Q(x), Q(y) \rangle_{X_{\mathbf{e}^{\dagger},\mathbf{i}}} \in \mathbb{C}(\mathbf{i}), \tag{4.27}$$

and

$$\langle x, y \rangle_{\mathbb{C}(\mathbf{j})} := \langle P(x) + Q(x), P(y) + Q(y) \rangle_{\mathbb{C}(\mathbf{j})}$$
$$:= \langle P(x), P(y) \rangle_{X_{\mathbf{e},\mathbf{j}}} + \langle Q(x), Q(y) \rangle_{X_{\mathbf{e}^{\dagger},\mathbf{j}}} \in \mathbb{C}(\mathbf{j}). \tag{4.28}$$

Of course, if both $(X_{\mathbf{e}}, \langle \cdot, \cdot \rangle_{X_{\mathbf{e},\mathbf{i}}})$ and $(X_{\mathbf{e}^{\dagger}}, \langle \cdot, \cdot \rangle_{X_{\mathbf{e}^{\dagger},\mathbf{i}}})$ are Hilbert spaces, then $(X_{\mathbb{C}(\mathbf{i})}, \langle \cdot, \cdot \rangle_{\mathbb{C}(\mathbf{i})})$ is also a Hilbert space. The same is true in the $\mathbb{C}(\mathbf{j})$ case.

Assume now that the bicomplex module X has a bicomplex inner product $\langle \cdot, \cdot \rangle$. By Theorem 4.5.2, $(X_{\mathbf{e}}, \langle \cdot, \cdot \rangle_{X_{\mathbf{e},\mathbf{i}}})$ and $(X_{\mathbf{e}^{\dagger}}, \langle \cdot, \cdot \rangle_{X_{\mathbf{e}^{\dagger},\mathbf{i}}})$ are $\mathbb{C}(\mathbf{i})$-inner product spaces, and $(X_{\mathbf{e}}, \langle \cdot, \cdot \rangle_{X_{\mathbf{e},\mathbf{j}}})$ and $(X_{\mathbf{e}^{\dagger}}, \langle \cdot, \cdot \rangle_{X_{\mathbf{e}^{\dagger},\mathbf{j}}})$ are $\mathbb{C}(\mathbf{j})$-inner product spaces, hence $X_{\mathbb{C}(\mathbf{i})}$

and $X_{\mathbb{C}(j)}$ are (complex) linear spaces with inner products $\langle \cdot, \cdot \rangle_{\mathbb{C}(i)}$ and $\langle \cdot, \cdot \rangle_{\mathbb{C}(j)}$ determined by Eqs. (4.27) and (4.28); moreover, $X_{\mathbb{C}(i)}$ and $X_{\mathbb{C}(j)}$ become (complex) normed spaces with the norms determined in the usual way by the formulas:

$$\|x\|_{\mathbb{C}(i)} := \langle x, x \rangle_{\mathbb{C}(i)}^{1/2}$$

$$:= \left(\langle P(x), P(x) \rangle_{X_{e,i}} + \langle Q(x), Q(x) \rangle_{X_{e^\dagger,i}} \right)^{1/2} \qquad (4.29)$$

$$= \left(\|P(x)\|_{X_{e,i}}^2 + \|Q(x)\|_{X_{e^\dagger,i}}^2 \right)^{1/2}$$

and similarly

$$\|x\|_{\mathbb{C}(j)} := \langle x, x \rangle_{\mathbb{C}(j)}^{1/2}$$

$$:= \left(\langle P(x), P(x) \rangle_{X_{e,j}} + \langle Q(x), Q(x) \rangle_{X_{e^\dagger,j}} \right)^{1/2} \qquad (4.30)$$

$$= \left(\|P(x)\|_{X_{e,j}}^2 + \|Q(x)\|_{X_{e^\dagger,j}}^2 \right)^{1/2}.$$

Recalling that

$$\mathbb{D}^+ \ni \langle x, x \rangle = \langle P(x), P(x) \rangle_{X_{e,i}} \mathbf{e} + \langle Q(x), Q(x) \rangle_{X_{e^\dagger,i}} \mathbf{e}^\dagger$$

$$= \langle P(x), P(x) \rangle_{X_{e,j}} \mathbf{e} + \langle Q(x), Q(x) \rangle_{X_{e^\dagger,j}} \mathbf{e}^\dagger,$$

one sees that
$$\langle P(x), P(x) \rangle_{X_{e,i}} = \langle P(x), P(x) \rangle_{X_{e,j}} \in \mathbb{R}^+ \cup \{0\}$$

and
$$\langle Q(x), Q(x) \rangle_{X_{e^\dagger,i}} = \langle Q(x), Q(x) \rangle_{X_{e^\dagger,j}} \in \mathbb{R}^+ \cup \{0\},$$

which implies
$$\|x\|_{\mathbb{C}(i)} = \|x\|_{\mathbb{C}(j)} \quad \forall x \in X. \qquad (4.31)$$

Formulas (1.8), (4.14), and the fact that we can take the square root of a bicomplex number by taking the square root of each of its idempotent components, imply that

$$\sqrt{\langle x, x \rangle} = \|P(x)\|_{X_{e,i}} \cdot \mathbf{e} + \|Q(x)\|_{X_{e^\dagger,i}} \cdot \mathbf{e}^\dagger$$

and therefore
$$\sqrt{2} |\sqrt{\langle x, x \rangle}| = \|x\|_{\mathbb{C}(i)} = \|x\|_{\mathbb{C}(j)} \quad \forall x \in X \qquad (4.32)$$

which shows the relation between both norms and the original bicomplex inner product.

Equation (4.31) tells us that although we have on X only one map

$$x \in X \mapsto \|x\|_{\mathbb{C}(\mathbf{i})} = \|x\|_{\mathbb{C}(\mathbf{j})}$$

such map determines two <u>different</u> norms on X seen as $X_{\mathbb{C}(\mathbf{i})}$ and $X_{\mathbb{C}(\mathbf{j})}$; in particular, X can be endowed with two linear topologies: one with scalars in $\mathbb{C}(\mathbf{i})$ and another with scalars in $\mathbb{C}(\mathbf{j})$. When we do not consider the field in which we take the scalars, we have the same underlying topology, which we will denote by $\tau_{\mathbb{C}}$; note that this topology is related to the initial bicomplex inner product via (4.32).

Note also that $|\sqrt{\langle x, x \rangle}|$ coincides with the norm (4.25). The approach given in Sect. 4.5.3 allows us to deal with the bicomplex norm on the bicomplex module X. What is more, the topology $\tau_{\mathbb{C}}$ is in fact a bicomplex linear topology, that is, the multiplication by bicomplex scalars (not only by $\mathbb{C}(\mathbf{i})$ scalars) is a continuous operation.

We conclude this section by illustrating some aspects of this (complex) approach to the notion of inner products on bicomplex modules using again $X = \mathbb{BC}$.

Note that formulas (1.4) and (1.6) give the following representations of a given bicomplex number:

$$
\begin{aligned}
Z &= (x_1 + \mathbf{i}\, y_1) + (x_2 + \mathbf{i}\, y_2)\mathbf{j} \\
&= (x_1 + \mathbf{j}\, x_2) + (y_1 + \mathbf{j}\, y_2)\mathbf{i} = \beta_1\, \mathbf{e} + \beta_2\, \mathbf{e}^\dagger \\
&:= ((x_1 + y_2) + \mathbf{i}\,(y_1 - x_2))\, \mathbf{e} + ((x_1 - y_2) + \mathbf{i}\,(y_1 + x_2))\, \mathbf{e}^\dagger \\
&= \gamma_1\, \mathbf{e} + \gamma_2\, \mathbf{e}^\dagger \\
&:= ((x_1 + y_2) + \mathbf{j}\,(x_2 - y_1))\, \mathbf{e} + ((x_1 - y_2) + \mathbf{j}\,(y_1 + x_2))\, \mathbf{e}^\dagger.
\end{aligned}
$$

Consider a second bicomplex number

$$
\begin{aligned}
W &= (s_1 + \mathbf{i}\, t_1) + (s_2 + \mathbf{i}\, t_2)\mathbf{j} \\
&= \xi_1\, \mathbf{e} + \xi_2\, \mathbf{e}^\dagger \\
&:= ((s_1 + t_2) + \mathbf{i}\,(t_1 - s_2))\, \mathbf{e} + ((s_1 - t_2) + \mathbf{i}\,(t_1 + s_2))\, \mathbf{e}^\dagger \\
&= \eta_1\, \mathbf{e} + \eta_2\, \mathbf{e}^\dagger \\
&:= ((s_1 + t_2) + \mathbf{j}\,(s_2 - t_1))\, \mathbf{e} + ((s_1 - t_2) + \mathbf{j}\,(t_1 + s_2))\, \mathbf{e}^\dagger.
\end{aligned}
$$

In terms of the previous sections, one has:

$$X = \mathbb{BC}; \quad X_{\mathbf{e},\mathbf{i}} = \mathbb{BC} \cdot \mathbf{e} = \mathbb{C}(\mathbf{i}) \cdot \mathbf{e}; \quad X_{\mathbf{e}^\dagger,\mathbf{i}} = \mathbb{BC} \cdot \mathbf{e}^\dagger = \mathbb{C}(\mathbf{i}) \cdot \mathbf{e}^\dagger;$$

$$\langle \beta_1 \, \mathbf{e}, \, \xi_1 \, \mathbf{e} \rangle_{\mathbb{C}(\mathbf{i}) \cdot \mathbf{e}} := \beta_1 \, \overline{\xi}_1 \, ;$$

$$\langle \beta_2 \, \mathbf{e}^\dagger, \, \xi_2 \, \mathbf{e}^\dagger \rangle_{\mathbb{C}(\mathbf{i}) \cdot \mathbf{e}^\dagger} := \beta_2 \, \overline{\xi}_2 \, ;$$

$$X_{\mathbf{e},\mathbf{j}} = \mathbb{BC} \cdot \mathbf{e} = \mathbb{C}(\mathbf{j}) \cdot \mathbf{e}; \quad X_{\mathbf{e}^\dagger,\mathbf{j}} = \mathbb{BC} \cdot \mathbf{e}^\dagger = \mathbb{C}(\mathbf{j}) \cdot \mathbf{e}^\dagger;$$

$$\langle \gamma_1 \, \mathbf{e}, \, \eta_1 \, \mathbf{e} \rangle_{\mathbb{C}(\mathbf{j}) \cdot \mathbf{e}} := \gamma_1 \, \eta_1^* \, ;$$

$$\langle \gamma_2 \, \mathbf{e}^\dagger, \, \eta_2 \, \mathbf{e}^\dagger \rangle_{\mathbb{C}(\mathbf{j}) \cdot \mathbf{e}^\dagger} := \gamma_2 \, \eta_2^* \, .$$

Then, formulas (4.27) and (4.28) give:

$$
\begin{aligned}
\langle Z, W \rangle_{\mathbb{C}(\mathbf{i})} &:= \langle ((x_1 + y_2) + \mathbf{i}\,(y_1 - x_2)) \, \mathbf{e}, \, ((s_1 + t_2) + \mathbf{i}\,(t_1 - s_2)) \, \mathbf{e} \rangle_{\mathbb{C}(\mathbf{i})\,\mathbf{e}} \\
&\quad + \langle ((x_1 - y_2) + \mathbf{i}\,(y_1 + x_2)) \, \mathbf{e}^\dagger, \, ((s_1 - t_2) + \mathbf{i}\,(t_1 + s_2)) \, \mathbf{e}^\dagger \rangle_{\mathbb{C}(\mathbf{i})\,\mathbf{e}^\dagger} \\
&= ((x_1 + y_2)\,(s_1 + t_2) + (y_1 - x_2)\,(t_1 - s_2) \\
&\quad + (x_1 - y_2)\,(s_1 - t_2) + (y_1 + x_2)\,(t_1 + s_2)) \\
&\quad + \mathbf{i}\,((y_1 - x_2)\,(s_1 + t_2) - (x_1 + y_2)\,(t_1 - s_2) \\
&\quad + (y_1 + x_2)\,(s_1 - t_2) - (x_1 - y_2)\,(t_1 + s_2)) \in \mathbb{C}(\mathbf{i}) \, ;
\end{aligned}
$$

$$
\begin{aligned}
\langle Z, W \rangle_{\mathbb{C}(\mathbf{j})} &:= \langle ((x_1 + y_2) + \mathbf{j}\,(x_2 - y_1)) \, \mathbf{e}, \, ((s_1 + t_2) + \mathbf{j}\,(s_2 - t_1)) \, \mathbf{e} \rangle_{\mathbb{C}(\mathbf{j})\,\mathbf{e}} \\
&\quad + \langle ((x_1 - y_2) + \mathbf{j}\,(y_1 + x_2)) \, \mathbf{e}^\dagger, \, ((s_1 - t_2) + \mathbf{j}\,(t_1 + s_2)) \, \mathbf{e}^\dagger \rangle_{\mathbb{C}(\mathbf{j})\,\mathbf{e}^\dagger} \\
&= ((x_1 + y_2)\,(s_1 + t_2) + (x_2 - y_1)\,(s_2 - t_1) \\
&\quad + (x_1 - y_2)\,(s_1 - t_2) + (y_1 + x_2)\,(t_1 + s_2)) \\
&\quad + \mathbf{j}\,((x_2 - y_1)\,(s_1 + t_2) - (x_1 + y_2)\,(s_2 - t_1) \\
&\quad + (y_1 + x_2)\,(s_1 - t_2) - (x_1 - y_2)\,(t_1 + s_2)) \in \mathbb{C}(\mathbf{j}) \, .
\end{aligned}
$$

Of course, $\langle Z, W \rangle_{\mathbb{C}(\mathbf{i})} \in \mathbb{C}(\mathbf{i})$ and $\langle Z, W \rangle_{\mathbb{C}(\mathbf{j})} \in \mathbb{C}(\mathbf{j})$ are different bicomplex numbers; but their "real parts" coincide, while their "imaginary parts" are in general different. They coincide, for example, when $Z = W$, in which case

$$\langle Z, Z \rangle_{\mathbb{C}(\mathrm{i})} = \langle Z, Z \rangle_{\mathbb{C}(\mathrm{j})} \geq 0 .$$

4.7 The Bicomplex Module $\mathbb{BC}^n$

4.7.1. We already know that $\mathbb{BC}^n$ is a bicomplex module. A bicomplex inner product on it can be introduced in a canonical way: if $Z = (Z_1, \ldots, Z_n)$, and $W = (W_1, \ldots, W_n) \in \mathbb{BC}^n$ then we set

$$\langle Z, W \rangle := Z_1 W_1^* + \cdots + Z_n W_n^* .$$

It is obvious that properties (1), (2), (3) of Definition 4.3.1 are satisfied. The last property can be proved similarly to the proof in Example 4.3.7, but we prefer to give an alternative proof using the idempotent representation. Indeed, write $Z_k = \beta_{k,1} \mathbf{e} + \beta_{k,2} \mathbf{e}^\dagger$ and correspondingly $Z_k^* = \overline{\beta}_{k,1} \mathbf{e} + \overline{\beta}_{k,2} \mathbf{e}^\dagger$. Then we have

$$\langle Z, Z \rangle = \sum_{k=1}^{n} Z_k \cdot Z_k^* = \sum_{k=1}^{n} \left(|\beta_{k,1}|^2 \mathbf{e} + |\beta_{k,2}|^2 \mathbf{e}^\dagger \right)$$

$$= \left(\sum_{k=1}^{n} |\beta_{k,1}|^2 \right) \mathbf{e} + \left(\sum_{k=1}^{n} |\beta_{k,2}|^2 \right) \mathbf{e}^\dagger \in \mathbb{D}^+ .$$

Note that the square of the inner product of some $Z \in \mathbb{BC}^n$ can be a zero divisor: this happens exactly when $Z \in \mathbb{BC}_{\mathbf{e}}^n = \mathbb{BC}^n \cdot \mathbf{e}$ or $Z \in \mathbb{BC}_{\mathbf{e}^\dagger}^n = \mathbb{BC}^n \cdot \mathbf{e}^\dagger$. Note also that this formula defines a hyperbolic norm on the $\mathbb{BC}$-module $\mathbb{BC}^n$.

4.7.2. Similar to what happens in the complex case, one can consider more general inner products on $\mathbb{BC}^n$. To do so, take any bicomplex positive matrix $A = \mathscr{A}_1 \mathbf{e} + \mathscr{A}_2 \mathbf{e}^\dagger$ and define, for $Z, W \in \mathbb{BC}^n$,

$$\langle Z, W \rangle_A := Z^t \cdot A \cdot W^* .$$

We can easily show that this is an inner product in the sense of Definition 4.3.1:

(1) $\langle Z, W + V \rangle_A = Z^t \cdot A \cdot (W + V)^* = Z^t \cdot A \cdot (W^* + V^*)$

$$= Z^t \cdot A \cdot W^* + Z^t \cdot A \cdot V^*$$

$$= \langle Z, W \rangle_A + \langle Z, V \rangle_A.$$

(2) Given $\mu \in \mathbb{BC}$, then

$$\langle \mu Z, W \rangle_A = (\mu Z)^t \cdot A \cdot W^* = \mu \left(Z^t \cdot A \cdot W^* \right)$$

$$= \mu \langle Z, W \rangle_A.$$

(3) $\langle Z, W \rangle_A = Z^t \cdot A \cdot W^* = \left(Z^t \cdot A \cdot W^* \right)^t$

$$= \left(\left(W^{*t} \cdot A^t \cdot Z \right)^* \right)^*$$

$$= \left(W^t \cdot A \cdot Z^* \right)^* = \langle W, Z \rangle_A^* ,$$

where we took into account that $Z^t \cdot A \cdot W^*$ is a 1×1 matrix.

(4) Since A is a bicomplex positive matrix, for any column $Z \in \mathbb{BC}^n$ we have that

$$Z^t \cdot A \cdot Z^* \in \mathbb{D}^+ .$$

Let us now reinterpret formula (4.24) in this setting. The simplest way to do this is via a direct computation. Take $Z = \mathscr{Z}_1 \mathbf{e} + \mathscr{Z}_2 \mathbf{e}^\dagger$, $W = \mathscr{W}_1 \mathbf{e} + \mathscr{W}_2 \mathbf{e}^\dagger \in \mathbb{BC}^n$ and let $A = \mathscr{A}_1 \mathbf{e} + \mathscr{A}_2 \mathbf{e}^\dagger$ be a bicomplex positive matrix. Then

$$\langle Z, W \rangle_A = \left(\mathscr{Z}_1 \mathbf{e} + \mathscr{Z}_2 \mathbf{e}^\dagger \right)^t \cdot \left(\mathscr{A}_1 \mathbf{e} + \mathscr{A}_2 \mathbf{e}^\dagger \right) \cdot \left(\mathscr{W}_1 \mathbf{e} + \mathscr{W}_2 \mathbf{e}^\dagger \right)^*$$

$$= \mathscr{Z}_1^t \cdot \mathscr{A}_1 \cdot \overline{\mathscr{W}}_1 \mathbf{e} + \mathscr{Z}_2^t \cdot \mathscr{A}_2 \cdot \overline{\mathscr{W}}_2 \mathbf{e}^\dagger$$

$$=: \langle \mathscr{Z}_1, \mathscr{W}_1 \rangle_{\mathscr{A}_1} \cdot \mathbf{e} + \langle \mathscr{Z}_2, \mathscr{W}_2 \rangle_{\mathscr{A}_2} \cdot \mathbf{e}^\dagger ,$$

which should be compared with formula (4.24).

The complex ($\mathbb{C}(\mathbf{i})$) matrices $\mathscr{A}_1$ and $\mathscr{A}_2$ are positive and thus $\langle \cdot, \cdot \rangle_{\mathscr{A}_1}$ and $\langle \cdot, \cdot \rangle_{\mathscr{A}_2}$ are ($\mathbb{C}(\mathbf{i})$) complex inner products on $\mathbb{C}^n(\mathbf{i})$. In particular,

$$\langle Z, Z \rangle_A = \langle \mathscr{Z}_1, \mathscr{Z}_1 \rangle_{\mathscr{A}_1} \cdot \mathbf{e} + \langle \mathscr{Z}_2, \mathscr{Z}_2 \rangle_{\mathscr{A}_2} \cdot \mathbf{e}^\dagger$$

$$= \| \mathscr{Z}_1 \|_{\mathbb{C}^n(\mathbf{i}), \mathscr{A}_1}^2 \cdot \mathbf{e} + \| \mathscr{Z}_2 \|_{\mathbb{C}^n(\mathbf{i}), \mathscr{A}_2}^2 \cdot \mathbf{e}^\dagger ,$$

which is the square of the hyperbolic norm on $\mathbb{BC}^n$. The bicomplex norm on $\mathbb{BC}^n$ generated by the inner product $\langle \cdot, \cdot \rangle_A$ is given by

$$\| Z \|_A := \frac{1}{\sqrt{2}} \sqrt{ \| \mathscr{Z}_1 \|_{\mathbb{C}^n(\mathbf{i}), \mathscr{A}_1}^2 + \| \mathscr{Z}_2 \|_{\mathbb{C}^n(\mathbf{i}), \mathscr{A}_2}^2 } .$$

4.8 The Ring $\mathbb{H}(\mathbb{C})$ of Biquaternions as a $\mathbb{BC}$-Module

Let us show now how this theory can clarify some algebraic properties of the ring of biquaternions (sometimes called complex quaternions). First of all recall that the ring $\mathbb{H}(\mathbb{C})$ of biquaternions is the set of elements of the form

$$\mathscr{Z} = z_0 + \mathbf{i}_1 z_1 + \mathbf{i}_2 z_2 + \mathbf{i}_2 z_2$$

with z_0, z_1, z_2, $z_3 \in \mathbb{C}(\mathbf{i})$; the three quaternionic imaginary units $\mathbf{i}_1$, $\mathbf{i}_2$, $\mathbf{i}_3$ are such that

$$\mathbf{i}_1^2 = \mathbf{i}_2^2 = \mathbf{i}_3^2 = -1,$$

the multiplication between them is anti-commutative:

$$\mathbf{i}_2 \cdot \mathbf{i}_1 = -\mathbf{i}_1 \cdot \mathbf{i}_2 = -\mathbf{i}_3; \quad \mathbf{i}_3 \cdot \mathbf{i}_2 = -\mathbf{i}_2 \cdot \mathbf{i}_3 = -\mathbf{i}_1;$$

$$\mathbf{i}_1 \cdot \mathbf{i}_3 = -\mathbf{i}_3 \cdot \mathbf{i}_1 = -\mathbf{i}_2,$$

while their products with the complex imaginary unit $\mathbf{i}$ are commutative:

$$\mathbf{i} \cdot \mathbf{i}_1 = \mathbf{i}_1 \cdot \mathbf{i}; \quad \mathbf{i} \cdot \mathbf{i}_2 = \mathbf{i}_2 \cdot \mathbf{i}; \quad \mathbf{i} \cdot \mathbf{i}_3 = \mathbf{i}_3 \cdot \mathbf{i}.$$

Since any of the products $\mathbf{i} \cdot \mathbf{i}_\ell$ satisfies $(\mathbf{i} \cdot \mathbf{i}_\ell)^2 = 1$, there are many hyperbolic units inside $\mathbb{H}(\mathbb{C})$ and thus there are many subsets inside it which are isomorphic to $\mathbb{BC}$, hence there are many different ways of making $\mathbb{H}(\mathbb{C})$ a $\mathbb{BC}$-module. Let us fix one of these ways and write each biquaternion as

$$\mathscr{L} = (z_0 + \mathbf{i}_1 z_1) + (z_2 + \mathbf{i}_1 z_2)\mathbf{i}_2 =: Z_1 + Z_2 \mathbf{i}_2,$$

where Z_1, $Z_2 \in \mathbb{BC}$, (i.e., our copy of $\mathbb{BC}$ has as imaginary units $\mathbf{i}$, $\mathbf{i}_1$, $\mathbf{k} = \mathbf{i}\mathbf{i}_1$). It is clear that $\mathbb{H}(\mathbb{C})$ is a $\mathbb{BC}$-module, with the multiplication by bicomplex scalars as follows. For any $\Lambda \in \mathbb{BC}$ and any $\mathscr{L} = Z_1 + Z_2 \mathbf{i}_2 \in \mathbb{H}(\mathbb{C})$ one has:

$$\Lambda \cdot \mathscr{L} = \Lambda \cdot Z_1 + \Lambda \cdot Z_2 \mathbf{i}_2.$$

Given $\mathscr{W} = W_1 + W_2 \mathbf{i}_2$ and taking into account that for any $Z \in \mathbb{BC}$ it is $\mathbf{i}_2 Z = Z^\dagger \mathbf{i}_2$, one has:

$$\mathscr{L} \cdot \mathscr{W} = \left(Z_1 W_1 - Z_2 W_2^\dagger\right) + \left(Z_1 W_2 + Z_2 W_1^\dagger\right) \mathbf{i}_2.$$

That is, $\mathbb{H}(\mathbb{C})$ becomes a ring and a $\mathbb{BC}$-module.

Because of the numerous imaginary units inside $\mathbb{H}(\mathbb{C})$, there are many conjugations, and they can be expressed using the conjugations from $\mathbb{BC}$. Some of them are (there is no standard notation in the literature):

$$\overline{\mathscr{L}} := (\bar{z}_0 + \bar{z}_1 \mathbf{i}_1) + (\bar{z}_2 + \bar{z}_3 \mathbf{i}_1)\mathbf{i}_2 = \overline{Z}_1 + \overline{Z}_2 \mathbf{i}_2 \,;$$

$$\mathscr{L}^{\dagger_1} := (z_0 - z_1 \mathbf{i}_1) + (z_2 - z_3 \mathbf{i}_1)\mathbf{i}_2 = Z_1^\dagger + Z_2^\dagger \mathbf{i}_2 \,;$$

$$\mathscr{L}^\star := (\bar{z}_0 - \bar{z}_1 \mathbf{i}_1) + (\bar{z}_2 - \bar{z}_3 \mathbf{i}_1)\mathbf{i}_2 = Z_1^* + Z_2^* \mathbf{i}_2 \,;$$

$$\mathscr{L}^{\dagger_2} := (z_0 + z_1 \mathbf{i}_1) - (z_2 + z_3 \mathbf{i}_1)\mathbf{i}_2 = Z_1 - Z_2 \mathbf{i}_2 \,;$$

$$\mathscr{Z}^{\dagger 3} := (z_0 + z_1 \mathbf{i}_1) + (z_2 - z_3 \mathbf{i}_1)\mathbf{i}_2 = Z_1 + Z_2^\dagger \mathbf{i}_2 ;$$

$$\left(\mathscr{Z}^{\dagger 1}\right)^{\dagger 2} = \left(\mathscr{Z}^{\dagger 2}\right)^{\dagger 1} = Z_1^\dagger - Z_2^\dagger \mathbf{i}_2 ;$$

$$\mathscr{Z}^\odot := Z_1^* - \overline{Z}_2 \mathbf{i}_2 ;$$

$$\mathscr{Z}^\diamond := Z_1^\dagger - Z_2 \mathbf{i}_2 ;$$

etc.

These conjugations interact with the product of two biquaternions as follows:

$$\overline{\mathscr{Z} \cdot \mathscr{W}} = \overline{\mathscr{Z}} \cdot \overline{\mathscr{W}} ; \qquad (\mathscr{Z} \cdot \mathscr{W})^{\dagger 1} = \mathscr{Z}^{\dagger 1} \cdot \mathscr{W}^{\dagger 1} ;$$

$$(\mathscr{Z} \cdot \mathscr{W})^\star = \mathscr{Z}^\star \cdot \mathscr{W}^\star ; \qquad (\mathscr{Z} \cdot \mathscr{W})^{\dagger 2} = \mathscr{Z}^{\dagger 2} \cdot \mathscr{W}^{\dagger 2} ;$$

$$(\mathscr{Z} \cdot \mathscr{W})^{\dagger 3} = \mathscr{W}^{\dagger 3} \cdot \mathscr{Z}^{\dagger 3} ; \qquad (\mathscr{Z} \cdot \mathscr{W})^\odot = \mathscr{W}^\odot \cdot \mathscr{Z}^\odot ;$$

$$(\mathscr{Z} \cdot \mathscr{W})^\diamond = \mathscr{W}^\diamond \cdot \mathscr{Z}^\diamond .$$

One may ask then if at least one of these conjugations is appropriate in order to generate an $\mathbb{H}(\mathbb{C})$-valued inner product with reasonably good properties. The answer is positive and is given as

Theorem 4.8.1 The mapping

$$\langle \cdot, \cdot \rangle_{\mathbb{H}(\mathbb{C})} : \mathbb{H}(\mathbb{C}) \times \mathbb{H}(\mathbb{C}) \longrightarrow \mathbb{H}(\mathbb{C})$$

given by

$$\langle \mathscr{Z}, \mathscr{W} \rangle_{\mathbb{H}(\mathbb{C})} := \mathscr{Z} \cdot \mathscr{W}^\odot ,$$

possesses the following properties:

   (I) $\langle \mathscr{Z}, \mathscr{W} + \mathscr{S} \rangle_{\mathbb{H}(\mathbb{C})} = \langle \mathscr{Z}, \mathscr{W} \rangle_{\mathbb{H}(\mathbb{C})} + \langle \mathscr{Z}, \mathscr{S} \rangle_{\mathbb{H}(\mathbb{C})};$
   (II) $\langle \Lambda \cdot \mathscr{Z}, \mathscr{W} \rangle_{\mathbb{H}(\mathbb{C})} = \Lambda \cdot \langle \mathscr{Z}, \mathscr{W} \rangle_{\mathbb{H}(\mathbb{C})};$
   (III) $\langle \mathscr{Z}, \mathscr{W} \rangle_{\mathbb{H}(\mathbb{C})} = \langle \mathscr{W}, \mathscr{Z} \rangle_{\mathbb{H}(\mathbb{C})}^\odot;$
   (IV) $\langle \mathscr{Z}, \mathscr{Z} \rangle_{\mathbb{H}(\mathbb{C})} = \eta + r\, \mathbf{i}\, \mathbf{i}_2 + s\, \mathbf{i}\, \mathbf{i}_3$ with $\eta \in \mathbb{D}^+, r, s \in \mathbb{R};$
       $\langle \mathscr{Z}, \mathscr{Z} \rangle_{\mathbb{H}(\mathbb{C})} = 0$ if and only if $\mathscr{Z} = 0.$

Proof:

The proof of (I), (II), and (III) follows by direct computation and taking into account that

$$\mathscr{Z} \cdot \mathscr{W}^\odot = (Z_1 + Z_2 \mathbf{i}_2) \cdot \left(W_1^* + \overline{W}_2 \mathbf{i}_2\right)$$

$$= \left(Z_1 W_1^* + Z_2 W_2^*\right) + \left(Z_2 \overline{W}_1 - Z_1 \overline{W}_2\right) \mathbf{i}_2 . \tag{4.33}$$

Since

$$\mathscr{Z} \cdot \mathscr{Z}^{\circ} = \left(Z_1 Z_1^* + Z_2 Z_2^*\right) + \left(Z_2 \overline{Z}_1 - Z_1 \overline{Z}_2\right) \mathbf{i}_2,$$

we may conclude that $\langle \mathscr{Z}, \mathscr{Z} \rangle_{\mathbb{H}(\mathbb{C})} = 0$ if and only if $\mathscr{Z} = 0$. In addition we know that $Z_1 Z_1^* + Z_2 Z_2^* \in \mathbb{D}^+$ and direct computations show that $\left(Z_2 \overline{Z}_1 - Z_1 \overline{Z}_2\right) \mathbf{i}_2$ is of the form $r \, \mathbf{i} \, \mathbf{i}_2 + s \, \mathbf{i} \, \mathbf{i}_3$ with $r, s \in \mathbb{R}$. $\qquad\square$

Corllary 4.8.2 For any $\mathscr{Z}$, $\mathscr{W}$ and Λ in $\mathbb{H}(\mathbb{C})$ it is:

$$\langle \mathscr{Z}, \Lambda \cdot \mathscr{W} \rangle_{\mathbb{H}(\mathbb{C})} = \langle \mathscr{Z}, \mathscr{W} \rangle_{\mathbb{H}(\mathbb{C})} \cdot \Lambda^{\circ}.$$

Thus, the biquaternionic module $\mathbb{H}(\mathbb{C})$ becomes endowed with an $\mathbb{H}(\mathbb{C})$-valued inner product with all the usual properties. Of course the inner product square is neither in $\mathbb{D}^+$ nor a real number but it has the peculiarity of involving the three hyperbolic imaginary units. We believe that in this way a theory of biquaternionic inner product spaces can be constructed along the same lines we have used to introduce the theory of bicomplex inner product modules.

Proposition 4.8.3 Let $\mathbb{H}(\mathbb{C})$ be seen as a $\mathbb{B}\mathbb{C}$-module. Then the first bicomplex component

$$Z_1 W_1^* + Z_2 W_2^*$$

of the biquaternionic inner product is a $\mathbb{B}\mathbb{C}$-valued inner product.

Proof:
This is an immediate consequence of the fact that we are considering $\mathbb{H}(\mathbb{C})$ as $\mathbb{B}\mathbb{C}^2$ and thus we can apply the definition and properties given in paragraph 4.7.1 $\qquad\square$

Some authors consider linear sets formed by $\mathbb{H}(\mathbb{C})$-valued functions endowing them with complex-valued inner products, but in this case some essential structures of such functions are lost. The Proposition above allows to catch more structures endowing such sets with bicomplex-valued inner products and consequenctly hyperbolic-valued norms.

Reference

1. R. Gervais Lavoie, L. Marchildon, D. Rochon, Infinite-dimensional bicomplex Hilbert spaces. Ann. Funct. Anal. **1**(2), 75–91 (2010)

Chapter 5
Linear Functionals and Linear Operators on $\mathbb{BC}$-Modules

Abstract This chapter treats, in the bicomplex context, the basic properties of linear functionals and linear operators. The notion of boundedness is given in terms of hyperbolic valued norms, thus allowing more similarity with the classical case. The bicomplex analogs of polarization identities are presented.

Keywords Linear functionals · Linear operators · Hyperbolic valued norms · $\mathbb{D}$-bounded operator · Polarization identities · Bicomplex modules.

5.1 Bicomplex Linear Functionals

5.1.1. Let X be a $\mathbb{BC}$-module, and let $f : X \to \mathbb{BC}$ be a $\mathbb{BC}$-linear functional. Then for any $x \in X$ one has:

$$f(x) = f_{1,\mathbf{i}}(x) \cdot \mathbf{e} + f_{2,\mathbf{i}}(x) \cdot \mathbf{e}^\dagger = f_{1,\mathbf{j}}(x) \cdot \mathbf{e} + f_{2,\mathbf{j}}(x) \cdot \mathbf{e}^\dagger$$

$$= F_1(x) + F_2(x) \cdot \mathbf{j} = G_1(x) + G_2(x) \cdot \mathbf{i} \in \mathbb{BC}$$

where $f_{\ell,\mathbf{i}}(x) \in \mathbb{C}(\mathbf{i})$, $f_{\ell,\mathbf{j}}(x) \in \mathbb{C}(\mathbf{j})$, $F_\ell(x) \in \mathbb{C}(\mathbf{i})$ and $G_\ell(x) \in \mathbb{C}(\mathbf{j})$, for $\ell = 1,\ 2$.
  This means that f induces maps

$$f_{1,\mathbf{i}},\ f_{2,\mathbf{i}} : X \to \mathbb{C}(\mathbf{i}), \qquad f_{1,\mathbf{j}},\ f_{2,\mathbf{j}} : X \to \mathbb{C}(\mathbf{j}),$$

$$F_1,\ F_2 : X \to \mathbb{C}(\mathbf{i}), \qquad G_1,\ G_2 : X \to \mathbb{C}(\mathbf{j}),$$

such that

$$f_{1,\mathbf{i}} = F_1 - \mathbf{i}\,F_2, \qquad f_{2,\mathbf{i}} = F_1 + \mathbf{i}\,F_2,$$

and

$$f_{1,\mathbf{j}} = G_1 - \mathbf{j}\,G_2, \qquad f_{2,\mathbf{j}} = G_1 + \mathbf{j}\,G_2.$$

D. Alpay et al., *Basics of Functional Analysis with Bicomplex Scalars, and Bicomplex Schur Analysis*, SpringerBriefs in Mathematics, DOI: 10.1007/978-3-319-05110-9_5, © The Author(s) 2014

In particular we have that

$$f_{\ell,\mathbf{i}} = \pi_{\ell,\mathbf{i}} \circ f = \left(\Pi_{1,\mathbf{i}} + (-1)^\ell \, \mathbf{i} \, \Pi_{2,\mathbf{i}}\right) \circ f;$$

$$f_{\ell,\mathbf{j}} = \pi_{\ell,\mathbf{j}} \circ f = \left(\Pi_{1,\mathbf{j}} + (-1)^\ell \, \mathbf{j} \, \Pi_{2,\mathbf{j}}\right) \circ f;$$

$$F_\ell = \Pi_{\ell,\mathbf{j}} \circ f = \frac{(\mathbf{j})^{\ell-1}}{2} \left(\pi_{1,\mathbf{j}} + (-1)^{\ell-1} \, \pi_{2,\mathbf{j}}\right) \circ f;$$

$$G_\ell = \Pi_{\ell,\mathbf{i}} \circ f = \frac{(\mathbf{i})^{\ell-1}}{2} \left(\pi_{1,\mathbf{i}} + (-1)^{\ell-1} \, \pi_{2,\mathbf{i}}\right) \circ f.$$

Let us show that $f_{1,\mathbf{i}}, \ f_{2,\mathbf{i}} : X_{\mathbb{C}(\mathbf{i})} \to \mathbb{C}(\mathbf{i})$ are $\mathbb{C}(\mathbf{i})$-linear functionals and that $f_{1,\mathbf{j}}, \ f_{2,\mathbf{j}} : X_{\mathbb{C}(\mathbf{j})} \to \mathbb{C}(\mathbf{j})$ are $\mathbb{C}(\mathbf{j})$-linear functionals. This is equivalent to requiring linearity for F_1, F_2, G_1 and G_2. Given $\lambda = \lambda_1 + \lambda_2\mathbf{j} \in \mathbb{BC}$, with $\lambda_1, \lambda_2 \in \mathbb{C}(\mathbf{i})$, and given $x, \ y \in X$, one has:

$$f(\lambda x + y) = f_{1,\mathbf{i}}(\lambda x + y)\mathbf{e} + f_{2,\mathbf{i}}(\lambda x + y)\mathbf{e}^\dagger$$

$$= \lambda f(x) + f(y)$$

$$= \left((\lambda_1 - \mathbf{i}\lambda_2)\, f_{1,\mathbf{i}}(x) + f_{1,\mathbf{i}}(y)\right)\mathbf{e}$$

$$+ \left((\lambda_1 + \mathbf{i}\lambda_2)\, f_{2,\mathbf{i}}(x) + f_{2,\mathbf{i}}(y)\right)\mathbf{e}^\dagger,$$

hence

$$f_{1,\mathbf{i}}(\lambda x + y) = (\lambda_1 - \mathbf{i}\lambda_2)\, f_{1,\mathbf{i}}(x) + f_{1,\mathbf{i}}(y) \tag{5.1}$$

and

$$f_{2,\mathbf{i}}(\lambda x + y) = (\lambda_1 - \mathbf{i}\lambda_2)\, f_{2,\mathbf{i}}(x) + f_{2,\mathbf{i}}(y). \tag{5.2}$$

In particular, setting $\lambda_2 = 0$ we have:

$$f_{1,\mathbf{i}}(\lambda_1 x + y) = \lambda_1 \, f_{1,\mathbf{i}}(x) + f_{1,\mathbf{i}}(y) \tag{5.3}$$

and

$$f_{2,\mathbf{i}}(\lambda_1 x + y) = \lambda_1 \, f_{2,\mathbf{i}}(x) + f_{2,\mathbf{i}}(y). \tag{5.4}$$

Thus, the mappings $f_{1,\mathbf{i}}, \ f_{2,\mathbf{i}}$ are $\mathbb{C}(\mathbf{i})$ linear functionals on $X_{\mathbb{C}(\mathbf{i})}$. The case of $f_{1,\mathbf{j}}$ and $f_{2,\mathbf{j}}$ follows in the same way.

In particular, Eqs. (5.1) and (5.2) show us the "type of homogeneity" of the $\mathbb{C}(\mathbf{i})$-linear functionals $f_{1,\mathbf{i}}$ and $f_{2,\mathbf{i}}$ with respect to general bicomplex scalars. Indeed, for $y = 0$ Eq. (5.1) gives:

$$f_{1,i}((\lambda_1 + \mathbf{j}\,\lambda_2)x) = (\lambda_1 - \mathbf{i}\,\lambda_2)\,f_{1,i}(x)\,,$$

that is,

$$f_{1,i}(\lambda x) = \pi_{1,i}(\lambda) \cdot f_{1,i}(x)\,. \tag{5.5}$$

In the same way Eq. (5.2) gives:

$$f_{2,i}(\lambda x) = \pi_{1,i}(\lambda) \cdot f_{2,i}(x)\,. \tag{5.6}$$

To understand this phenomenon we note that if $\lambda = \lambda_1 \mathbf{e} + \lambda_2 \mathbf{e}^\dagger \in \mathbb{BC}$, and $x \in X$, then

$$f(\lambda x) = f_{1,i}(\lambda x)\mathbf{e} + f_{2,i}(\lambda x)\mathbf{e}^\dagger$$

$$= \pi_{1,i} \circ f(\lambda x)\,\mathbf{e} + \pi_{2,i} \circ f(\lambda x)\,\mathbf{e}^\dagger$$

$$= \pi_{1,i}(\lambda\,f(x))\,\mathbf{e} + \pi_{2,i}(\lambda\,f(x))\,\mathbf{e}^\dagger$$

$$= \pi_{1,i}(\lambda)\,\pi_{1,i}(f(x))\,\mathbf{e} + \pi_{2,i}(\lambda)\,\pi_{2,i}(f(x))\,\mathbf{e}^\dagger$$

$$= \pi_{1,i}(\lambda)\,f_{1,i}(x)\,\mathbf{e} + \pi_{2,i}(\lambda)\,f_{2,i}(x)\,\mathbf{e}^\dagger\,.$$

Thus, if $\lambda = \lambda_1 \in \mathbb{C}(\mathbf{i})$, we obtain the $\mathbb{C}(\mathbf{i})$-linearity of $f_{1,i}$ and $f_{2,i}$. In general when $\lambda \in \mathbb{BC}$ we have (5.5) and (5.6).

5.1.2. As in the classic complex case the components of a bicomplex linear functional f are not independent. For example, one has that

$$f(\mathbf{j}x) = F_1(\mathbf{j}x) + \mathbf{j}\,F_2(\mathbf{j}x)$$
$$= \mathbf{j}\,f(x) = -F_2(x) + \mathbf{j}\,F_1(x)$$

which implies

$$F_1(x) = F_2(\mathbf{j}x)\,,$$
$$F_2(x) = -F_1(\mathbf{j}x)\,,$$

which are of course mutually reciprocal relations. In the same way

$$G_1(x) = G_2(\mathbf{i}x)\,,$$
$$G_2(x) = -G_1(\mathbf{i}x)\,.$$

This phenomenon is even more remarkable for the functionals $f_{1,i}$, $f_{2,i}$, $f_{1,j}$ and $f_{2,j}$. Indeed

$$f_{1,\mathbf{i}}(\mathbf{j}\,x) = F_1(\mathbf{j}\,x) - \mathbf{i}\,F_2(\mathbf{j}\,x) = -F_2(x) - \mathbf{i}\,F_1(x)$$
$$= -\mathbf{i}\,(F_1(x) - \mathbf{i}\,F_2(x)) = -\mathbf{i}\,f_{1,\mathbf{i}}(x)$$

and

$$f_{2,\mathbf{i}}(\mathbf{j}\,x) = F_1(\mathbf{j}\,x) + \mathbf{i}\,F_2(\mathbf{j}\,x) = -F_2(x) + \mathbf{i}\,F_1(x)$$
$$= \mathbf{i}\,(F_1(x) + \mathbf{i}\,F_2(x)) = \mathbf{i}\,f_{2,\mathbf{i}}(x)\,,$$

that is,

$$f_{1,\mathbf{i}}(\mathbf{j}\,x) = -\mathbf{i}\,f_{1,\mathbf{i}}(x),$$
$$f_{2,\mathbf{i}}(\mathbf{j}\,x) = \mathbf{i}\,f_{2,\mathbf{i}}(x)\,.$$

In the same way

$$f_{1,\mathbf{j}}(\mathbf{i}\,x) = -\mathbf{j}\,f_{1,\mathbf{j}}(x)\,,$$
$$f_{2,\mathbf{j}}(\mathbf{i}\,x) = \mathbf{j}\,f_{2,\mathbf{j}}(x)\,.$$

Consider now a bicomplex linear functional $f : X \to \mathbb{BC}$. One has that $f(x) = f(\mathbf{e}\,x + \mathbf{e}^\dagger\,x) = f(x\,\mathbf{e})\,\mathbf{e} + f(x\,\mathbf{e}^\dagger)\,\mathbf{e}^\dagger = f_{1,\mathbf{i}}(x\,\mathbf{e})\,\mathbf{e} + f_{2,\mathbf{i}}(x\,\mathbf{e}^\dagger)\,\mathbf{e}^\dagger = f_{1,\mathbf{j}}(x\,\mathbf{e})\,\mathbf{e} + f_{2,\mathbf{j}}(x\,\mathbf{e}^\dagger)\,\mathbf{e}^\dagger$, i. e.,

$$f(x) = f_{1,\mathbf{i}}(x\,\mathbf{e})\,\mathbf{e} + f_{2,\mathbf{i}}(x\,\mathbf{e}^\dagger)\,\mathbf{e}^\dagger$$
$$= f_{1,\mathbf{j}}(x\,\mathbf{e})\,\mathbf{e} + f_{2,\mathbf{j}}(x\,\mathbf{e}^\dagger)\,\mathbf{e}^\dagger\,. \tag{5.7}$$

Let us assume now that X has a real-valued or a hyperbolic norm. Then equalities (5.7) imply that f is continuous if and only if $f_{\ell,\mathbf{i}}$, $f_{\ell,\mathbf{j}}$, $\ell = 1, 2$, are continuous on $X_{\mathbb{C}(\mathbf{i})}$ and on $X_{\mathbb{C}(\mathbf{j})}$ respectively. In particular if f is continuous, then the restrictions $f_{1,\mathbf{i}}\,|_{X_{\mathbf{e},\mathbf{i}}}$ and $f_{2,\mathbf{i}}\,|_{X_{\mathbf{e}^\dagger,\mathbf{i}}}$ are $\mathbb{C}(\mathbf{i})$-linear continuous functionals; in the same manner $f_{1,\mathbf{j}}\,|_{X_{\mathbf{e},\mathbf{j}}}$ and $f_{2,\mathbf{j}}\,|_{X_{\mathbf{e}^\dagger,\mathbf{j}}}$ are $\mathbb{C}(\mathbf{j})$-linear continuous functionals.

We are now ready to prove a bicomplex version of the celebrated Riesz theorem (technically, such theorem was already proved in [1], but we decided to include its proof here for the sake of completeness, and because we can frame the proof in the context of our more general treatment of bicomplex functionals).

Theorem 5.1.3 (Riesz). Let X be a bicomplex Hilbert space and let $f : X \to \mathbb{BC}$ be a continuous $\mathbb{BC}$-linear functional. Then there exists a unique $y \in X$ such that for every $x \in X$, $f(x) = \langle x, y \rangle$.

Proof:
The first representation in (5.7) generates two $\mathbb{C}(\mathbf{i})$-linear continuous functionals on $X_{\mathbf{e},\mathbf{i}}$ and $X_{\mathbf{e}^\dagger,\mathbf{i}}$ for each of which the complex Ricsz representation Theorem gives a unique element $u = \mathbf{e}\,u \in X_{\mathbf{e},\mathbf{i}}$ and $v = \mathbf{e}^\dagger\,v \in X_{\mathbf{e}^\dagger,\mathbf{i}}$ such that

$$f_{1,\mathbf{i}}\,|_{X_{\mathbf{e},\mathbf{i}}}\,(\mathbf{e}\,x) = \langle \mathbf{e}\,x, \mathbf{e}\,u \rangle_{X_{\mathbf{e},\mathbf{i}}} \tag{5.8}$$

and

$$f_{2,i} \, |_{X_{\mathbf{e}^\dagger,i}} \, (\mathbf{e}^\dagger x) = \langle \mathbf{e}^\dagger x, \mathbf{e}^\dagger v \rangle_{X_{\mathbf{e}^\dagger,i}} . \tag{5.9}$$

The obvious candidate $y := u + v = \mathbf{e}u + \mathbf{e}^\dagger v \in X$ allows us to conclude the proof since

$$
\begin{aligned}
\langle y, x \rangle &= \langle \mathbf{e}\,x + \mathbf{e}^\dagger x, \, \mathbf{e}\,u + \mathbf{e}^\dagger v \rangle \\
&= \mathbf{e}\,\langle \mathbf{e}\,x, \mathbf{e}\,u \rangle_{X_{\mathbf{e},i}} + \mathbf{e}^\dagger \langle \mathbf{e}^\dagger x, \, \mathbf{e}^\dagger v \rangle_{X_{\mathbf{e}^\dagger,i}} \\
&= \mathbf{e}\, f_{1,i}(\mathbf{e}\,x) + \mathbf{e}^\dagger \, f_{2,i}(\mathbf{e}^\dagger x) = f(x).
\end{aligned}
$$

Since the bicomplex inner product is non degenerate then the element y is unique. At the same time this unicity implies that if we repeat the above process using the $\mathbb{C}(\mathbf{j})$-linear spaces $X_{\mathbf{e},\mathbf{j}}$ and $X_{\mathbf{e}^\dagger,\mathbf{j}}$ the element that we would get will be the same. $\square$

The next result can also be found in [1].

Theorem 5.1.4 (First bicomplex Schwarz inequality). Let X be a bicomplex Hilbert module, and let $x, y \in X$. Then

$$|\langle x, y \rangle| \leq \sqrt{2}\,\|x\|\,\|y\|.$$

Proof:
Recall that X is the direct sum of $X_{\mathbf{e}}$ and $X_{\mathbf{e}^\dagger}$ when they are seen as $\mathbb{C}(\mathbf{i})$- or $\mathbb{C}(\mathbf{j})$-complex Hilbert spaces. For any $x, y \in X$ one has

$$
\begin{aligned}
|\langle x, y \rangle| &= \left| \langle \mathbf{e}\,x + \mathbf{e}^\dagger x, \, \mathbf{e}\,y + \mathbf{e}^\dagger y \rangle \right| \\
&= \left| \mathbf{e}\,\langle \mathbf{e}\,x, \mathbf{e}\,y \rangle_{X_{\mathbf{e},i}} + \mathbf{e}^\dagger \langle \mathbf{e}^\dagger x, \, \mathbf{e}^\dagger y \rangle_{X_{\mathbf{e}^\dagger,i}} \right|
\end{aligned}
$$

using (1.8)

$$= \frac{1}{\sqrt{2}} \left(\left| \langle \mathbf{e}\,x, \mathbf{e}\,y \rangle_{X_{\mathbf{e},i}} \right|^2 + \left| \langle \mathbf{e}^\dagger x, \, \mathbf{e}^\dagger y \rangle_{X_{\mathbf{e}^\dagger,i}} \right|^2 \right)^{1/2}$$

using the complex Schwarz inequality

$$
\begin{aligned}
&\leq \frac{1}{\sqrt{2}} \left(\|\mathbf{e}\,x\|_{X_{\mathbf{e},i}}^2 \cdot \|\mathbf{e}\,y\|_{X_{\mathbf{e},i}}^2 + \left\|\mathbf{e}^\dagger x\right\|_{X_{\mathbf{e}^\dagger,i}}^2 \cdot \left\|\mathbf{e}^\dagger y\right\|_{X_{\mathbf{e}^\dagger,i}}^2 \right)^{1/2} \\
&\leq \frac{1}{\sqrt{2}} \left(2\,\|x\|^2 \left(\|\mathbf{e}\,y\|_{X_{\mathbf{e},i}}^2 + \left\|\mathbf{e}^\dagger y\right\|_{X_{\mathbf{e}^\dagger,i}}^2 \right) \right)^{1/2} \\
&= \left(2\,\|x\|^2 \cdot \|y\|^2 \right)^{1/2} = \sqrt{2}\,\|x\| \cdot \|y\|,
\end{aligned}
$$

which concludes the proof. $\square$

Since we have a hyperbolic norm which takes values in the partially ordered set $\mathbb{D}^+$, we are able to prove a "hyperbolic version" of the Schwarz inequality.

Theorem 5.1.5 (Second bicomplex Schwarz inequality). Let X be a bicomplex Hilbert module, and let $x, y \in X$. Then

$$|\langle x, y \rangle|_{\mathbf{k}} \preccurlyeq \|x\|_{\mathbb{D}} \cdot \|y\|_{\mathbb{D}} .$$

Proof:

As before,

$$\langle x, y \rangle = \mathbf{e} \langle \mathbf{e}x, \mathbf{e}y \rangle_{X_{\mathbf{e}}} + \mathbf{e}^{\dagger} \langle \mathbf{e}^{\dagger} x, \mathbf{e}^{\dagger} y \rangle_{X_{\mathbf{e}^{\dagger}}} ,$$

hence

$$|\langle x, y \rangle|_{\mathbf{k}} = \mathbf{e} |\langle \mathbf{e}x, \mathbf{e}y \rangle_{X_{\mathbf{e}}}| + \mathbf{e}^{\dagger} \cdot |\langle \mathbf{e}^{\dagger}x, \mathbf{e}^{\dagger}y \rangle_{X_{\mathbf{e}^{\dagger}}}|$$

using the complex Schwarz inequality

$$\begin{aligned}
&\preccurlyeq \mathbf{e} \|\mathbf{e}x\|_{X_{\mathbf{e}}} \cdot \|\mathbf{e}y\|_{X_{\mathbf{e}}} + \mathbf{e}^{\dagger} \|\mathbf{e}^{\dagger}x\|_{X_{\mathbf{e}^{\dagger}}} \cdot \|\mathbf{e}^{\dagger}y\|_{X_{\mathbf{e}^{\dagger}}} \\
&= \left(\mathbf{e} \cdot \|\mathbf{e}x\|_{X_{\mathbf{e}}} + \mathbf{e}^{\dagger} \|\mathbf{e}^{\dagger}x\|_{X_{\mathbf{e}^{\dagger}}} \right) \cdot \left(\mathbf{e} \cdot \|\mathbf{e}y\|_{X_{\mathbf{e}}} + \mathbf{e}^{\dagger} \|\mathbf{e}^{\dagger}y\|_{X_{\mathbf{e}^{\dagger}}} \right) \\
&= \|x\|_{\mathbb{D}} \cdot \|y\|_{\mathbb{D}}.
\end{aligned}$$
□

Notice the absence of the coefficient $\sqrt{2}$ in the second inequality; this is due to the compatibility between the hyperbolic modulus of the inner product (which is a bicomplex number) and the hyperbolic norm on X.

Example 5.1.6 We are interested again in the situation in which $X = \mathbb{BC}$. Take then $x = Z = \beta_1 \mathbf{e} + \beta_2 \mathbf{e}^{\dagger}$, and $y = W = \xi_1 \mathbf{e} + \xi_2 \mathbf{e}^{\dagger}$, $\beta_1, \beta_2, \xi_1, \xi_2 \in \mathbb{C}(\mathbf{i})$ so that

$$\|Z\|_X^2 = \frac{1}{2} \left(|\beta_1|^2 + |\beta_2|^2 \right) = |Z|^2, \quad \|W\|_X^2 = \frac{1}{2} \left(|\xi_1|^2 + |\xi_2|^2 \right) = |W|^2.$$

One has directly:

$$\begin{aligned}
|\langle Z, W \rangle| = |Z \cdot W^*| &= |(\beta_1 \mathbf{e} + \beta_2 \mathbf{e}^{\dagger})(\overline{\xi}_1 \mathbf{e} + \overline{\xi}_2 \mathbf{e}^{\dagger})| \\
&= |\beta_1 \overline{\xi}_1 \mathbf{e} + \beta_2 \overline{\xi}_2 \mathbf{e}^{\dagger}| \\
&= \frac{1}{\sqrt{2}} \sqrt{|\beta_1|^2 \cdot |\overline{\xi}_1|^2 + |\beta_2|^2 \cdot |\overline{\xi}_2|^2} \\
&\leq \frac{1}{\sqrt{2}} \sqrt{2 \|Z\|_{\mathbb{C}}^2 \left(|\xi_1|^2 + |\xi_2|^2 \right)} \\
&= \sqrt{2 \|Z\|_X^2 \cdot \|W\|_X^2} = \sqrt{2} \|Z\|_X \cdot \|W\|_X ,
\end{aligned}$$

in accordance with Theorem 5.1.4.

Similarly, for the hyperbolic modulus of the inner product and for the corresponding hyperbolic norm one has:

$$\begin{aligned}
|\langle Z, W \rangle|_{\mathbf{k}} = |Z \cdot W^*|_{\mathbf{k}} &= |Z|_{\mathbf{k}} \cdot |W^*|_{\mathbf{k}} \\
&= |Z|_{\mathbf{k}} \cdot |W|_{\mathbf{k}} = \|Z\|_{\mathbb{D}} \cdot \|W\|_{\mathbb{D}} ;
\end{aligned}$$

This means that, in this case, the Second Schwarz inequality becomes an equality, exactly as it happens with the Schwarz inequality in $\mathbb{C}$.

5.2 Polarization Identities

Let X be a real linear space with a bilinear symmetric form $\mathcal{B}_{\mathbb{R}}(\cdot, \cdot)$, denote by $\mathcal{Q}_{\mathbb{R}}(\cdot)$ the corresponding quadratic form generated by $\mathcal{B}_{\mathbb{R}}$, i.e., for any $x \in X$,

$$\mathcal{Q}_{\mathbb{R}}(x) := \mathcal{B}_{\mathbb{R}}(x, x).$$

Then the following relation holds for any $x, y \in X$:

$$\mathcal{B}_{\mathbb{R}}(x, y) = \frac{1}{4} \left(\mathcal{Q}_{\mathbb{R}}(x + y) - \mathcal{Q}_{\mathbb{R}}(x - y) \right). \tag{5.10}$$

This relationship usually goes under the name of Polarization Identity for real linear spaces. In particular, a real inner product and its associated norm satisfy (5.10).

For a complex linear space X the form $\mathcal{B}_{\mathbb{C}}$ is assumed to be Hermitian and sesquilinear, that is, $\mathcal{B}_{\mathbb{C}}(x, y) = \overline{\mathcal{B}_{\mathbb{C}}(y, x)}$ and $\mathcal{B}_{\mathbb{C}}(\lambda x, \mu y) = \lambda \overline{\mu} \mathcal{B}_{\mathbb{C}}(x, y)$ for any $x, y \in X$, and for all $\lambda, \mu \in \mathbb{C}$. let $\mathcal{Q}_{\mathbb{C}}(x) := \mathcal{B}_{\mathbb{C}}(x, x)$ denote the quadratic form. The polarization identity in this case becomes

$$\mathcal{B}_{\mathbb{C}}(x, y) = \frac{1}{4} \left(\mathcal{Q}_{\mathbb{C}}(x + y) - \mathcal{Q}_{\mathbb{C}}(x - y) \right.$$
$$\left. + \mathbf{i} \left(\mathcal{Q}_{\mathbb{C}}(x + \mathbf{i} y) - \mathcal{Q}_{\mathbb{C}}(x - \mathbf{i} y) \right) \right). \tag{5.11}$$

Again any complex inner product and its corresponding norm satisfy (5.11).

Let now X be a bicomplex module. We will consider maps

$$\mathcal{B} : X \times X \to \mathbb{BC}$$

such that $\mathcal{B}(x, y_1 + y_2) = \mathcal{B}(x, y_1) + \mathcal{B}(x, y_2)$ and $\mathcal{B}(x_1 + x_2, y) = \mathcal{B}(x_1, y) + \mathcal{B}(x_2, y)$ for all $x_1, x_2, x, y_1, y_2, y \in X$. For such a map $\mathcal{B}$, to be "Hermitian" may mean one of the three properties:

(I) If $\mathcal{B}(x, y) = \overline{\mathcal{B}(y, x)}$, then $\mathcal{B}$ is called bar-Hermitian,

(II) If $\mathcal{B}(x, y) = (\mathcal{B}(y, x))^{\dagger}$, then $\mathcal{B}$ is called $\dagger$-Hermitian,

(III) If $\mathcal{B}(x, y) = (\mathcal{B}(y, x))^{*}$, then $\mathcal{B}$ is called $*$-Hermitian.

Similarly we have the three types of sesquilinearity. If $\nu, \mu \in \mathbb{BC}$ then

(I') $\mathcal{B}$ is bar-sesquilinear if

$$\mathcal{B}(\nu x, \mu y) = \nu \overline{\mu} \, \mathcal{B}(x, y).$$

(II') $\mathcal{B}$ is called †-sesquilinear if

$$\mathcal{B}(\nu x, \mu y) = \nu \mu^\dagger \, \mathcal{B}(x, y).$$

(III') $\mathcal{B}$ is ∗-sesquilinear if

$$\mathcal{B}(\nu x, \mu y) = \nu \mu^* \, \mathcal{B}(x, y).$$

For reasons that will be readily apparent, we are going to consider ∗-Hermitian and ∗-sesquilinear maps $\mathcal{B}$. Note that since $\mathcal{Q}(x) := \mathcal{B}(x, x) = \mathcal{B}(x, x)^*$ then $\mathcal{Q}(x)$ takes values in $\mathbb{D}$. Hence the inner products that we have defined fit into this notion: if $\mathcal{B}(x, y) = \langle x, y \rangle_X$ then $\mathcal{B}$ is a ∗-Hermitian and ∗-sesquilinear form, and $\mathcal{Q}(x) = \mathcal{B}(x, x)$ is the square of a hyperbolic norm on X.

It turns out that given a ∗-Hermitian and ∗-sesquilinear bicomplex form $\mathcal{B}$, there are several polarization-type identities that connect it to its quadratic form $\mathcal{Q}$. Indeed, it is a simple and direct computation to show that for any $x, y \in X$ one has

$$
\begin{aligned}
\mathcal{B}(x, y) = \frac{1}{4} \, (&\mathcal{Q}(x + y) - \mathcal{Q}(x - y) \\
&+ \mathbf{i}\,(\mathcal{Q}(x + \mathbf{i}y) - \mathcal{Q}(x - \mathbf{i}y)));
\end{aligned}
\tag{5.12}
$$

$$
\begin{aligned}
\mathcal{B}(x, y) = \frac{1}{4} \, (&\mathbf{j}\,(\mathcal{Q}(x + \mathbf{j}y) - \mathcal{Q}(x - \mathbf{j}y)) \\
&+ \mathbf{k}\,(\mathcal{Q}(x + \mathbf{k}y) - \mathcal{Q}(x - \mathbf{k}y)));
\end{aligned}
\tag{5.13}
$$

$$
\begin{aligned}
\mathcal{B}(x, y) = \frac{1}{4} \, (&\mathcal{Q}(x + y) - \mathcal{Q}(x - y) \\
&+ \mathbf{j}\,(\mathcal{Q}(x + \mathbf{j}y) - \mathcal{Q}(x - \mathbf{j}y)));
\end{aligned}
\tag{5.14}
$$

$$
\begin{aligned}
\mathcal{B}(x, y) = \frac{1}{4} \, (&\mathbf{i}\,(\mathcal{Q}(x + \mathbf{i}y) - \mathcal{Q}(x - \mathbf{i}y)) \\
&+ \mathbf{k}\,(\mathcal{Q}(x + \mathbf{k}y) - \mathcal{Q}(x - \mathbf{k}y))).
\end{aligned}
\tag{5.15}
$$

The validity of formulas (5.12) and (5.14) can be concluded from (5.11) but there are bicomplex reasons for the validity of (5.12)–(5.15). Indeed, $\mathcal{B}(x, y)$ can be written as

$$
\begin{aligned}
\mathcal{B}(x, y) &= \mathcal{B}(x\mathbf{e} + x\mathbf{e}^\dagger, \, y\mathbf{e} + y\mathbf{e}^\dagger) \\
&= \mathbf{e}\,\mathcal{B}(x\mathbf{e}, y\mathbf{e}) + \mathbf{e}^\dagger \mathcal{B}(x\mathbf{e}^\dagger, y\mathbf{e}^\dagger) \\
&= \mathbf{e}\,\pi_{1,\mathbf{i}}(\mathcal{B}(x\mathbf{e}, y\mathbf{e})) + \mathbf{e}^\dagger \pi_{2,\mathbf{i}}(\mathcal{B}(x\mathbf{e}^\dagger, y\mathbf{e}^\dagger)) \\
&= \mathbf{e}\,\pi_{1,\mathbf{j}}(\mathcal{B}(x\mathbf{e}, y\mathbf{e})) + \mathbf{e}^\dagger \pi_{2,\mathbf{j}}(\mathcal{B}(x\mathbf{e}^\dagger, y\mathbf{e}^\dagger)),
\end{aligned}
\tag{5.16}
$$

so that

$$
\begin{aligned}
\mathcal{Q}(x) &= \mathcal{B}(x, x) \\
&= \mathbf{e}\,\pi_{1,\mathbf{i}}(\mathcal{B}(x\mathbf{e}, y\mathbf{e})) + \mathbf{e}^\dagger \pi_{2,\mathbf{i}}(\mathcal{B}(x\mathbf{e}^\dagger, y\mathbf{e}^\dagger)) \\
&= \mathbf{e}\,\pi_{1,\mathbf{i}}(\mathcal{Q}(x\mathbf{e})) + \mathbf{e}^\dagger \pi_{2,\mathbf{i}}(\mathcal{Q}(x\mathbf{e}^\dagger)) \\
&= \mathbf{e}\,\pi_{1,\mathbf{j}}(\mathcal{B}(x\mathbf{e}, y\mathbf{e})) + \mathbf{e}^\dagger \pi_{2,\mathbf{j}}(\mathcal{B}(x\mathbf{e}^\dagger, y\mathbf{e}^\dagger)) \\
&= \mathbf{e}\,\pi_{1,\mathbf{j}}(\mathcal{Q}(x\mathbf{e})) + \mathbf{e}^\dagger \pi_{2,\mathbf{j}}(\mathcal{Q}(x\mathbf{e}^\dagger)).
\end{aligned}
$$

By applying this to the right-hand side of (5.13), we see that for the first summand one has

$$
\begin{aligned}
\mathbf{j}\,(Q(x+\mathbf{j}y) &- Q(x-\mathbf{j}y)) \\
&= \mathbf{j}\left(\mathbf{e}\,\pi_{1,i}\left(Q((x+\mathbf{j}y)\mathbf{e})\right) + \mathbf{e}^\dagger\pi_{2,i}\left(Q((x+\mathbf{j}y)\mathbf{e}^\dagger)\right)\right. \\
&\quad \left. - \mathbf{e}\,\pi_{1,i}\left(Q((x-\mathbf{j}y)\mathbf{e})\right) - \mathbf{e}^\dagger\pi_{2,i}\left(Q((x-\mathbf{j}y)\mathbf{e}^\dagger)\right)\right) \\
&= -\mathbf{i}\mathbf{e}\,\pi_{1,i}\left(Q((x-\mathbf{i}y)\mathbf{e})\right) + \mathbf{i}\mathbf{e}^\dagger\pi_{2,i}\left(Q((x+\mathbf{i}y)\mathbf{e}^\dagger)\right) \\
&\quad + \mathbf{i}\mathbf{e}\,\pi_{1,i}\left(Q((x+\mathbf{i}y)\mathbf{e})\right) - \mathbf{i}\mathbf{e}^\dagger\pi_{2,i}\left(Q((x-\mathbf{i}y)\mathbf{e}^\dagger)\right) \\
&= \mathbf{i}\left(\mathbf{e}\,\pi_{1,i}\left(Q((x+\mathbf{i}y)\mathbf{e})\right) + \mathbf{e}^\dagger\pi_{2,i}\left(Q((x+\mathbf{i}y)\mathbf{e}^\dagger)\right)\right. \\
&\quad \left. - \left(\mathbf{e}\,\pi_{1,i}\left(Q((x-\mathbf{i}y)\mathbf{e})\right) + \mathbf{e}^\dagger\pi_{2,i}\left(Q((x-\mathbf{i}y)\mathbf{e}^\dagger)\right)\right)\right) \\
&= \mathbf{i}\,(Q(x+\mathbf{i}y) - Q(x-\mathbf{i}y));
\end{aligned}
$$

as to the second summand we have:

$$
\begin{aligned}
\mathbf{k}\,(Q(x+\mathbf{k}y) &- Q(x-\mathbf{k}y)) \\
&= \mathbf{k}\left(\mathbf{e}\,\pi_{1,i}\left(Q((x+\mathbf{k}y)\mathbf{e})\right) + \mathbf{e}^\dagger\pi_{2,i}\left(Q((x+\mathbf{k}y)\mathbf{e}^\dagger)\right)\right. \\
&\quad \left. - \mathbf{e}\,\pi_{1,i}\left(Q((x-\mathbf{k}y)\mathbf{e})\right) - \mathbf{e}^\dagger\pi_{2,i}\left(Q((x-\mathbf{k}y)\mathbf{e}^\dagger)\right)\right) \\
&= \mathbf{e}\,\pi_{1,i}\left(Q((x+y)\mathbf{e})\right) - \mathbf{e}^\dagger\pi_{2,i}\left(Q((x-y)\mathbf{e}^\dagger)\right) \\
&\quad - \mathbf{e}\,\pi_{1,i}\left(Q((x-y)\mathbf{e})\right) + \mathbf{e}^\dagger\pi_{2,i}\left(Q((x+y)\mathbf{e}^\dagger)\right) \\
&= Q(x+y) - Q(x-y).
\end{aligned}
$$

One can now compare the right-hand sides in (5.12)–(5.15) to conclude that they all coincide.

Take a bicomplex module X. As we saw in (5.16), any Hermitian $*$-sesquilinear form $\mathcal{B}$ has "an idempotent representation"

$$
\begin{aligned}
\mathcal{B}(x, y) &= \mathbf{e}\,\mathcal{B}(\mathbf{e}x, \mathbf{e}y) + \mathbf{e}^\dagger\mathcal{B}(\mathbf{e}^\dagger x, \mathbf{e}^\dagger y) \\
&= \mathbf{e}\,\pi_{1,i}(\mathcal{B}(\mathbf{e}x, \mathbf{e}y)) + \mathbf{e}^\dagger\,\pi_{2,i}(\mathcal{B}(\mathbf{e}^\dagger x, \mathbf{e}^\dagger y)).
\end{aligned}
$$

On $X_\mathbf{e} = \mathbf{e}X$ set

$$
\mathfrak{b}_{1,i}(\mathbf{e}x, \mathbf{e}y) := \pi_{1,i}(\mathcal{B}(\mathbf{e}x, \mathbf{e}y)).
$$

We will show that $\mathfrak{b}_{1,i}$ is a Hermitian $\mathbb{C}(\mathbf{i})$-sesquilinear form on $X_{\mathbf{e},i}$. Indeed,

$$
ll(1)\ \overline{\mathfrak{b}_{1,i}(\mathbf{e}y, \mathbf{e}x)} = \overline{\pi_{1,i}(\mathcal{B}(\mathbf{e}y, \mathbf{e}x))}
$$

$$
= \overline{\pi_{1,i}(\mathcal{B}(\mathbf{e}x, \mathbf{e}y)^*)}
$$

$$
= \overline{\pi_{1,i}\left(\left(\mathbf{e}\,\pi_{1,i}(\mathcal{B}(\mathbf{e}x, \mathbf{e}y)) + \mathbf{e}^\dagger\,\pi_{2,i}(\mathcal{B}(\mathbf{e}^\dagger x, \mathbf{e}^\dagger y))\right)^*\right)}
$$

$$= \pi_{1,i}\left(e\,\overline{\pi_{1,i}(\mathcal{B}(ex, ey))} + e^\dagger\,\overline{\pi_{2,i}(\mathcal{B}(e^\dagger x, e^\dagger y))}\right)$$

$$= \overline{\overline{\pi_{1,i}(\mathcal{B}(ex, ey))}} = \pi_{1,i}(\mathcal{B}(ex, ey))$$

$$= \mathfrak{b}_{1,i}(ex, ey).$$

$$(2)\ \mathfrak{b}_{1,i}(\mu\,ex + ez, ey) = \pi_{1,i}(\mathcal{B}(\mu\,ex + ez, ey))$$

$$= \pi_{1,i}(\mu\,\mathcal{B}(ex, ey) + \mathcal{B}(ez, ey))$$

$$= \mu\,\mathfrak{b}_{1,i}(ex, ey) + \mathfrak{b}_{1,i}(ez, ey).$$

The two properties together imply that

$$\mathfrak{b}_{1,i}(ex, \mu\,ey) = \overline{\mu}\,\mathfrak{b}_{1,i}(ex, ey).$$

In the same way we can define three more forms

$$\mathfrak{b}_{2,i}(e^\dagger x, e^\dagger y) := \pi_{2,i}(\mathcal{B}(e^\dagger x, e^\dagger y)),$$
$$\mathfrak{b}_{1,j}(ex, ey) := \pi_{1,j}(\mathcal{B}(ex, ey)),$$
$$\mathfrak{b}_{2,j}(e^\dagger x, e^\dagger y) := \pi_{2,j}(\mathcal{B}(e^\dagger x, e^\dagger y)),$$

which turn out to be Hermitian complex sesquilinear forms on $X_{e^\dagger,i}$, $X_{e,j}$, $X_{e^\dagger,j}$ respectively.

One can also obtain polarization identities for matrices, which directly descend from those we have described above. Indeed, any matrix $A \in \mathbb{BC}^{n \times n}$ generates two $*$-Hermitian matrices

$$A_1 := A + A^{*t} \qquad \text{and} \qquad A_2 := \mathbf{i}\,(A - A^{*t}).$$

It is easy to see that the formulas

$$\mathcal{B}_1(x,\ y) := x^t \cdot A_1 \cdot y^*$$

and

$$\mathcal{B}_2(x,\ y) := x^t \cdot A_2 \cdot y^*$$

define $*$-Hermitian and $*$-sesquilinear bicomplex forms on the bicomplex module $\mathbb{BC}^n$, with $\mathcal{Q}_1(x) := \mathcal{B}_1(x,\ x) = x^t \cdot A_1 \cdot x^*$ and $\mathcal{Q}_2(x) := \mathcal{B}_2(x,\ x) = x^t \cdot A_2 \cdot x^*$. Thus any of the formulas (5.12)–(5.15) implies the corresponding formula for $\mathcal{B}_1$,

$\mathcal{Q}_1$ and $\mathcal{B}_2$, $\mathcal{Q}_2$. For instance, formula (5.12) applied to the bicomplex form $\mathcal{B}_2$ gives, for any matrix A,

$$
\begin{aligned}
x^t \cdot &\left(A - A^{*\,t}\right) \cdot y^* \\
&= \frac{1}{4} \left((x+y)^t \cdot (A - A^{*\,t}) \cdot (x+y)^* \right. \\
&\quad - (x-y)^t \cdot (A - A^{*\,t}) \cdot (x-y)^* \\
&\quad \left. + \mathbf{i} \left((x+\mathbf{i}y)^t \cdot (A - A^{*\,t}) \cdot (x+\mathbf{i}y)^* - (x-\mathbf{i}y)^t \cdot (A - A^{*\,t}) \cdot (x-\mathbf{i}y)^* \right) \right).
\end{aligned}
$$
(5.17)

Assume additionally that for any column $b \in \mathbb{BC}^n$ the matrix A verifies:

$$
b^t \cdot A \cdot b^* \in \mathbb{D};
$$

the hyperbolic numbers seen as 1×1 matrices are invariant under transposition and under $*$-involution, hence

$$
b^t \cdot A \cdot b^* = (b^t \cdot A \cdot b^*)^{*\,t} = (b^{*\,t} \cdot A^* \cdot b)^t = b^t \cdot A^{*\,t} \cdot b^*,
$$

which implies the vanishing of the right-hand side of (5.17) for such a matrix. Thus

$$
x^t \cdot (A - A^{*\,t}) \cdot y^* = 0
$$

for any x, y and so

$$
A = A^{*\,t}.
$$

5.3 Linear Operators on $\mathbb{BC}$-Modules

Let X and Y be two $\mathbb{BC}$-modules. A map $T : X \to Y$ is said to be a linear operator if for any x, $z \in X$ and for any $\lambda \in \mathbb{BC}$ it is

$$
T[\lambda x + z] = \lambda T[x] + T[z].
$$

Defining as usual the sum of two operators and the multiplication of a linear operator by a bicomplex number, one can see that the set of all linear operators forms a $\mathbb{BC}$-module. In case when $Y = X$ the same set equipped additionally with the composition becomes a non-commutative ring and a bicomplex algebra.

Let us now see how the idempotent decompositions of X and Y manifest themselves in the properties of a linear operator. Write $X = X_{\mathbf{e}} + X_{\mathbf{e}^\dagger}$, $Y = Y_{\mathbf{e}} + Y_{\mathbf{e}^\dagger}$ so that any $x \in X$ is of the form

$$
x = \mathbf{e} \cdot \mathbf{e}x + \mathbf{e}^\dagger \cdot \mathbf{e}^\dagger x =: \mathbf{e} \cdot x_1 + \mathbf{e}^\dagger \cdot x_2.
$$

The linearity of T gives:

$$T[x] = T[\mathbf{e} \cdot \mathbf{e}x + \mathbf{e}^\dagger \cdot \mathbf{e}^\dagger x]$$

$$= \mathbf{e} \cdot T[\mathbf{e}x] + \mathbf{e}^\dagger \cdot T[\mathbf{e}^\dagger x]$$

$$= \mathbf{e} \cdot (\mathbf{e} \cdot T[\mathbf{e}x]) + \mathbf{e}^\dagger \cdot \left(\mathbf{e}^\dagger \cdot T[\mathbf{e}^\dagger x]\right),$$

and introducing the operators T_1 and T_2 by

$$T_1[x] := \mathbf{e} \cdot T[\mathbf{e}x], \qquad T_2[x] := \mathbf{e}^\dagger \cdot T[\mathbf{e}^\dagger x]$$

we obtain what can be called the idempotent representation of a linear operator:

$$T = \mathbf{e} \cdot T_1 + \mathbf{e}^\dagger \cdot T_2 = T_1 + T_2.$$

Although T_1 and T_2 are defined on the whole X their ranges do not fill all of Y: the range $T_1(X)$ lies inside $Y_\mathbf{e}$ and the range $T_2(X)$ lies inside $Y_{\mathbf{e}^\dagger}$; in addition, the action of T_1 is determined by the component $x_1 = \mathbf{e}x$ of x and the action of T_2 is determined by the component $x_2 = \mathbf{e}^\dagger x$ of x. Thus, with a somewhat imprecise terminology, we can say that T_1 is an operator from $X_\mathbf{e}$ into $Y_\mathbf{e}$, and T_2 is an operator from $X_{\mathbf{e}^\dagger}$ into $Y_{\mathbf{e}^\dagger}$. Since T_1 and T_2 are obviously $\mathbb{BC}$-linear then one may consider them as $\mathbb{C}(\mathbf{i})$- or $\mathbb{C}(\mathbf{j})$-linear on $X_\mathbf{e}$ and $X_{\mathbf{e}^\dagger}$ respectively.

Let $T : X \longrightarrow Y$ be a $\mathbb{BC}$-linear operator. Assume additionally that X and Y have $\mathbb{D}$-norms $\| \cdot \|_{\mathbb{D}, X}$ and $\| \cdot \|_{\mathbb{D}, Y}$ respectively (as well as the real valued norms $\| \cdot \|_X$ and $\| \cdot \|_Y$). Recall that we can always write $X = X_\mathbf{e} + X_{\mathbf{e}^\dagger}$, $Y = Y_\mathbf{e} + Y_{\mathbf{e}^\dagger}$, that the norms $\| \cdot \|_{X_\mathbf{e}}$, $\| \cdot \|_{X_{\mathbf{e}^\dagger}}$ are the restrictions of $\| \cdot \|_X$ to $X_\mathbf{e}$ and to $X_{\mathbf{e}^\dagger}$, and that $\| \cdot \|_{Y_\mathbf{e}}$, $\| \cdot \|_{Y_{\mathbf{e}^\dagger}}$ are the restrictions of $\| \cdot \|_Y$ to $Y_\mathbf{e}$ and to $Y_{\mathbf{e}^\dagger}$.

Definition 5.3.1 The operator $T : X \longrightarrow Y$ is called $\mathbb{D}$-bounded if there exists $\Lambda \in \mathbb{D}^+$ such that for any $x \in X$ one has

$$\|Tx\|_{\mathbb{D}, Y} \preccurlyeq \Lambda \cdot \|x\|_{\mathbb{D}, X}. \tag{5.18}$$

The $\mathbb{D}$-infimum of these Λ is called the norm of the operator T.

It is clear that the set of $\mathbb{D}$-bounded $\mathbb{BC}$ operators is a $\mathbb{BC}$-module. Indeed, let T_1 and T_2 be two $\mathbb{D}$-bounded operators:

$$\|T_1 x\|_{\mathbb{D}, Y} \preccurlyeq \Lambda_1 \cdot \|x\|_{\mathbb{D}, X}, \qquad \|T_2 x\|_{\mathbb{D}, Y} \preccurlyeq \Lambda_2 \cdot \|x\|_{\mathbb{D}, X} \forall x \in X,$$

then we have

$$\|(T_1 + T_2)[x]\|_{\mathbb{D}, Y} \preccurlyeq \|T_1 x\|_{\mathbb{D}, Y} + \|T_2 x\|_{\mathbb{D}, Y}$$

$$\preccurlyeq \Lambda_1 \cdot \|x\|_{\mathbb{D}, X} + \Lambda_2 \cdot \|x\|_{\mathbb{D}, X}.$$

If Λ_1 and Λ_2 are comparable with respect to the partial order $\preccurlyeq$ then

$$\| (T_1 + T_2) [x] \|_{\mathbb{D},Y} \preccurlyeq \max\{\Lambda_1, \Lambda_2\} \cdot \|x\|_{\mathbb{D},X}.$$

But the hyperbolic numbers Λ_1 and Λ_2 may be non comparable. In this case the following bound works. If $\Lambda_1 = a_1 \mathbf{e} + b_1 \mathbf{e}^\dagger$, $\Lambda_2 = a_2 \mathbf{e} + b_2 \mathbf{e}^\dagger$ then set $a := \max\{a_1, a_2\}$, $b := \max\{b_1, b_2\}$, and $\Lambda := a\mathbf{e} + b\mathbf{e}^\dagger$ which gives

$$\| (T_1 + T_2) [x] \|_{\mathbb{D},Y} \preccurlyeq \Lambda \cdot \|x\|_{\mathbb{D},X} \quad \forall x \in X,$$

thus, $T_1 + T_2$ is $\mathbb{D}$-bounded. The case of the multiplication by bicomplex scalars is even simpler since

$$\|\mu T x\|_{\mathbb{D},Y} = \|\mu\|_{\mathbf{k}} \cdot \|Tx\|_{\mathbb{D},X} \preccurlyeq \|\mu\|_{\mathbf{k}} \cdot \Lambda \cdot \|x\|_{\mathbb{D},X} \quad \forall x \in X.$$

One can finally prove, following the usual argument from classical functional analysis, that the map

$$T \mapsto \inf_{\mathbb{D}} \{\Lambda \mid \Lambda \text{ satisfies eq. } (5.18)\} =: \|T\|_{\mathbb{D}}$$

defines a hyperbolic norm on the $\mathbb{BC}$-module of all bounded operators.

Note that inequality (5.18) is equivalent to asking that, for any $x \in X$,

$$\|Tx\|_{\mathbb{D},Y} = \|T_1 x_1 + T_2 x_2\|_{\mathbb{D},Y}$$

$$= \mathbf{e} \cdot \|T_1 x_1\|_{Y_\mathbf{e}} + \mathbf{e}^\dagger \cdot \|T_2 x_2\|_{Y_\mathbf{e}^\dagger}$$

$$\preccurlyeq \mathbf{e} \cdot \Lambda_1 \cdot \|x_1\|_{X_\mathbf{e}} + \mathbf{e}^\dagger \cdot \Lambda_2 \cdot \|x_2\|_{X_\mathbf{e}^\dagger},$$

with $\Lambda = \mathbf{e}\Lambda_1 + \mathbf{e}^\dagger \Lambda_2 \in \mathbb{D}^+$. This means in particular that T_1 and T_2 are bounded and $\|T_1\| = \inf\{\Lambda_1\}$, $\|T_2\| = \inf\{\Lambda_2\}$. It is clear now that if T_1 and T_2 are bounded then T is $\mathbb{D}$-bounded. Thus, the operator T is $\mathbb{D}$-bounded if and only if the operators $T_1 : X_\mathbf{e} \to Y_\mathbf{e}$ and $T_2 : X_\mathbf{e}^\dagger \to Y_\mathbf{e}^\dagger$ are bounded. What is more, the respective norms are connected by equation

$$\|T\|_{\mathbb{D}} = \mathbf{e} \cdot \|T_1\| + \mathbf{e}^\dagger \cdot \|T_2\|, \tag{5.19}$$

which is true because

$$\inf_{\mathbb{D}}\{\Lambda\} = \inf_{\mathbb{D}}\{\mathbf{e}\Lambda_1 + \mathbf{e}^\dagger \Lambda_2\}$$

$$= \mathbf{e} \cdot \inf\{\Lambda_1\} + \mathbf{e}^\dagger \cdot \inf\{\Lambda_2\} = \mathbf{e} \cdot \|T_1\| + \mathbf{e}^\dagger \|T_2\|.$$

As always, the operator T is (sequentially) continuous at a point $x \in X$ if for any $\{x_n\} \subset X$ such that $\lim_{n \to \infty} \|x_n - x\|_{\mathbb{D},X} = 0$ it holds that $\lim_{n \to \infty} \|T x_n - T x\|_{\mathbb{D},Y} = 0$. It is obvious that T is continuous at x if and only if T_1 is continuous at $\mathbf{e}x$ and T_2 is continuous at $\mathbf{e}^\dagger x$. This implies immediately that T is $\mathbb{D}$-bounded if and only if T is continuous.

In the classical case, the norms of bounded operators on complex spaces can be represented in different ways:

$$\|T_1\| = \sup\left\{ \frac{\|T_1 z\|_{Y_\mathbf{e}}}{\|z\|_{X_\mathbf{e}}} \mid \|z\|_{X_\mathbf{e}} \neq 0 \right\}$$

$$= \sup\left\{ \|T_1 z\|_{Y_\mathbf{e}} \mid \|z\|_{X_\mathbf{e}} \leq 1 \right\} \tag{5.20}$$

$$= \sup\left\{ \|T_1 z\|_{Y_\mathbf{e}} \mid \|z\|_{X_\mathbf{e}} = 1 \right\};$$

(and similarly for T_2). By using these equalities, formula (5.19) gives

$$\|T\|_{\mathbb{D}} = \mathbf{e} \cdot \sup\left\{ \frac{\|T_1 z\|_{Y_\mathbf{e}}}{\|z\|_{X_\mathbf{e}}} \mid z \in X_\mathbf{e} \setminus \{0\} \right\}$$

$$+ \mathbf{e}^\dagger \cdot \sup\left\{ \frac{\|T_2 u\|_{Y_\mathbf{e}^\dagger}}{\|u\|_{X_\mathbf{e}^\dagger}} \mid u \in X_\mathbf{e}^\dagger \setminus \{0\} \right\}$$

$$= \sup_{\mathbb{D}} \left\{ \frac{\mathbf{e} \cdot \|T_1 z\|_{Y_\mathbf{e}} + \mathbf{e}^\dagger \cdot \|T_2 u\|_{Y_\mathbf{e}^\dagger}}{\mathbf{e} \cdot \|z\|_{X_\mathbf{e}} + \mathbf{e}^\dagger \cdot \|u\|_{X_\mathbf{e}^\dagger}} \mid \mathbf{e}\|z\|_{X_\mathbf{e}} + \mathbf{e}^\dagger\|u\|_{X_\mathbf{e}^\dagger} \notin \mathfrak{S}_0 \right\}$$

$$= \sup_{\mathbb{D}} \left\{ \frac{\|\mathbf{e} \cdot T_1 z + \mathbf{e}^\dagger \cdot T_2 u\|_{\mathbb{D}, Y}}{\|\mathbf{e} \cdot z + \mathbf{e}^\dagger \cdot u\|_{\mathbb{D}, X}} \mid \|\mathbf{e}z + \mathbf{e}^\dagger u\|_{\mathbb{D}, X} \notin \mathfrak{S}_0 \right\}$$

$$= \sup_{\mathbb{D}} \left\{ \frac{\|Tw\|_{\mathbb{D}, Y}}{\|w\|_{\mathbb{D}, X}} \mid \|w\|_{\mathbb{D}, X} \notin \mathfrak{S}_0 \right\},$$

that is, we have proved for $\mathbb{D}$-bounded operators an exact analog of the first formula in (5.20).

Similarly, we can prove the analogs of the other two formulas in (5.20).

$$\|T\|_{\mathbb{D}} = \sup_{\mathbb{D}} \left\{ \|Tw\|_{\mathbb{D}, Y} \mid \|w\|_{\mathbb{D}, X} \prec 1 \right\}$$

$$= \sup_{\mathbb{D}} \left\{ \|Tw\|_{\mathbb{D}, Y} \mid \|w\|_{\mathbb{D}, X} = 1 \right\}.$$

Assume now that X and Y are bicomplex Hilbert modules with inner products $\langle \cdot, \cdot \rangle_X$ and $\langle \cdot, \cdot \rangle_Y$ respectively. Recall that, with obvious meanings of the symbols,

$$\langle x, z \rangle_X = \mathbf{e} \cdot \langle \mathbf{e}x, \mathbf{e}z \rangle_{X_\mathbf{e}} + \mathbf{e}^\dagger \cdot \langle \mathbf{e}^\dagger x, \mathbf{e}^\dagger z \rangle_{X_{\mathbf{e}^\dagger}}.$$

The adjoint operator $T^\sharp : Y \to X$ for a bounded operator $T : Y \to X$ is defined by the equality

$$\langle T[x], y \rangle_Y = \langle x, T^\sharp[y] \rangle_X.$$

Consider the left-hand side:

$$
\begin{aligned}
\langle T[x], y \rangle_Y &= \langle (T_1 + T_2) \left[\mathbf{e}x + \mathbf{e}^\dagger x \right], \mathbf{e}y + \mathbf{e}^\dagger y \rangle_Y \\
&= \langle \mathbf{e} \cdot T_1[\mathbf{e}x] + \mathbf{e}^\dagger \cdot T_2[\mathbf{e}^\dagger x], \mathbf{e}y + \mathbf{e}^\dagger y \rangle_Y \\
&= \mathbf{e} \cdot \langle T_1[\mathbf{e}x], \mathbf{e}y \rangle_{Y_\mathbf{e}} + \mathbf{e}^\dagger \cdot \langle T_2[\mathbf{e}^\dagger x], \mathbf{e}^\dagger y \rangle_{Y_{\mathbf{e}^\dagger}} \\
&= \mathbf{e} \cdot \langle \mathbf{e}x, T_1^\sharp[\mathbf{e}y] \rangle_{Y_\mathbf{e}} + \mathbf{e}^\dagger \cdot \langle \mathbf{e}^\dagger x, T_2^\sharp[\mathbf{e}^\dagger y] \rangle_{Y_{\mathbf{e}^\dagger}};
\end{aligned}
$$

similarly for the right-hand side:

$$
\begin{aligned}
\langle x, T^\sharp[y] \rangle_X &= \langle \mathbf{e}x + \mathbf{e}^\dagger x, \left((T^\sharp)_1 + (T^\sharp)_2 \right) [\mathbf{e}y + \mathbf{e}^\dagger y] \rangle_X \\
&= \mathbf{e} \cdot \langle \mathbf{e}x, (T^\sharp)_1[\mathbf{e}y] \rangle_X + \mathbf{e}^\dagger \cdot \langle \mathbf{e}^\dagger x, (T^\sharp)_2[\mathbf{e}^\dagger y] \rangle_X.
\end{aligned}
$$

Thus, one concludes that the bicomplex adjoint $T^\sharp$ always exists and its idempotent components $(T^\sharp)_1$ and $(T^\sharp)_2$ are the complex adjoints of the operators T_1 and T_2, that is,

$$T^\sharp = \mathbf{e} \cdot T_1^\sharp + \mathbf{e}^\dagger \cdot T_2^\sharp. \tag{5.21}$$

Reference

1. R. Gervais Lavoie, L. Marchildon, D. Rochon, Infinite-dimensional bicomplex Hilbert spaces. Ann. Funct. Anal. **1**(2), 75–91 (2010)

Chapter 6
Schur Analysis

Abstract This last chapter is an application of the previous results, and it begins with a study of Schur analysis in the bicomplex setting. Further results, and in particular the associated theory of linear systems, will be presented elsewhere.

Keywords Blaschke factors · Hardy spaces · Positive definite functions · Schur analysis · Schur multipliers

6.1 A Survey of Classical Schur Analysis

Under the term *Schur analysis* one understands problems associated to the class of functions s that are holomorphic and whose modulus is bounded by one in the open unit disk $\mathbb{K}$ (we will call such functions Schur functions, or Schur multipliers, and denote their class by $\mathscr{S}$). They can be interpreted as the transfer functions of time-invariant dissipative linear systems and, as such, also play an important role in the theory of linear systems (see [1] for a survey). It was in the early years of the twentieth century that I. Schur wrote two particular works [2, 3] in 1917 and 1918; they were motivated by the works of Carathéodory and Fejér [4], Herglotz [5] and Toeplitz [6] (in particular in the trigonometric moment problem). In those works Schur associated to a Schur function s a sequence $\rho_0, \rho_1, \ldots$ of complex numbers in the open unit disk: one sets $s_0(z) = s(z)$, $\rho_0 = s_0(0)$; then one defines recursively a sequence of Schur functions $s_0, s_1, \ldots$ as follows: if $|s_n(0)| = 1$ the process ends. If $|s_n(0)| < 1$, one defines

$$
s_{n+1}(z) = \begin{cases} \dfrac{s_n(z) - s_n(0)}{z(1 - \overline{s_n(0)}s_n(z))}, & \text{if } z \neq 0, \\[3mm] \dfrac{s_n'(0)}{(1 - |s_n(0)|^2)}, & \text{if } z = 0. \end{cases}
$$

D. Alpay et al., *Basics of Functional Analysis with Bicomplex Scalars,
and Bicomplex Schur Analysis*, SpringerBriefs in Mathematics,
DOI: 10.1007/978-3-319-05110-9_6, © The Author(s) 2014

Then s_{n+1} is still a Schur function in view of Schwartz' lemma, and one sets $\rho_{n+1} := s_{n+1}(0)$. The coefficients $\rho_0, \rho_1, \ldots$ are called the Schur coefficients of s and they characterize in a unique way the function s. The construction of the sequence $s_0, s_1, \ldots$ is called Schur's algorithm. Since the function s is holomorphic, it admits its Taylor expansion:

$$s(z) = a_0 + a_1 z + a_2 z^2 + \cdots + a_n z^n + \cdots,$$

and the sequence of coefficients $\{a_n\}$ also characterize in a unique way the function s. It is remarkable that the sequence of Schur's coefficients $\{\rho_n\}$ has proved to give a better characterization of s and makes possible to link such functions with, for instance, problems of interpolation of holomorphic functions, filtering of stationary stochastic processes, etc. Examples of Schur functions are the so-called Blaschke factors:

$$b_\omega(z) = \frac{z - \omega}{1 - z\overline{\omega}},$$

with $\omega \in \mathbb{K}$, and more general finite products:

$$s(z) = c \cdot \prod_{\ell=1}^{k} b_{\omega_\ell}, \tag{6.1}$$

where c is a complex number with modulus one and $\omega_1, \omega_2, \ldots, \omega_k$ are in $\mathbb{K}$. I. Schur proved that the sequence of coefficients $\{\rho_n\}$ is finite if and only if s is a function of the type (6.1), i.e., if s is a finite Blaschke product.

The role of Schur functions in the theory of time-invariant dissipative linear systems has led to the extension of the definition of Schur functions to other settings. For instance, the case of several complex variables corresponds to multi-indexed (ND) systems (recall that a linear system can be seen as a holomorphic or a rational function defined in the unit disk and whose values are matrices). Analogs of Schur functions were introduced by Agler, see [7, 8]. Besides the Schur-Agler classes, we mention that counterparts of Schur functions have been studied in the time-varying case, [9], the stochastic case, [10], the Riemann surface case, [11], and the multiscale case, [12, 13]. A number of other cases also are of importance. In the setting of hypercomplex analysis, counterparts of Schur functions were studied in [14–17] in the setting of Fueter series, and in [18] in the setting of slice hyperholomorphic functions. In this chapter we study the analog of Schur functions in the setting of bicomplex numbers and $\mathbb{BC}$-holomorphic functions.

We begin with the following characterization of Schur functions of one complex variable.

Theorem 6.1.1 Let s be a function holomorphic in the unit disk $\mathbb{K}$. The following are equivalent:

(i) s is a Schur function.
(ii) The kernel

$$k_s(\lambda, \mu) = \frac{1 - s(\lambda)\,\overline{s(\mu)}}{1 - \lambda\,\overline{\mu}} \tag{6.2}$$

is positive definite in $\mathbb{K}$, that is, given any collection $\lambda_1, \dots, \lambda_n$ of points in $\mathbb{K}$, the matrix $(k_s(\lambda_\ell, \lambda_m))_{\ell, m}$ is positive definite.
(iii) One can write s in the form

$$s(\lambda) = D + \lambda C(I_{\mathcal{H}(s)} - \lambda A)^{-1}B, \tag{6.3}$$

where $\mathcal{H}(s)$ is a Hilbert space and where the operator matrix

$$\begin{pmatrix} A & B \\ C & D \end{pmatrix} \;:\; \mathcal{H}(s) \oplus \mathbb{C} \quad \longrightarrow \quad \mathcal{H}(s) \oplus \mathbb{C}$$

is coisometric (i.e., $\begin{pmatrix} A & B \\ C & D \end{pmatrix} \cdot \begin{pmatrix} A & B \\ C & D \end{pmatrix}^{\sharp} = I_{\mathcal{H}(s) \oplus \mathbb{C}}$, where we are using the symbol $\sharp$ instead of $*$ to denote the dual of the operator matrix) or unitary.

(iv) The multiplication operator M_s defined by $M_s[f] := sf$ is a contraction from the Hardy space $\mathbf{H}^2(\mathbb{K})$ into itself.

Expression (6.3) is called a realization of s. Here we need to recall that a Hilbert space $\mathcal{H}$ whose elements are functions defined on a set Ω and with values in $\mathbb{C}^n$ is called a reproducing kernel Hilbert space if for any $c \in \mathbb{C}^n$ and any $w \in \Omega$ the functional $f \mapsto \overline{c}^t f(w)$ is bounded; by Riesz' Theorem (see for instance [19] Theorem 4.12, p. 77) there exists a function

$$K(z, w) : \Omega \times \Omega \to \mathbb{C}^{n \times n}$$

such that for any $c \in \mathbb{C}^n$, $w \in \Omega$ and $f \in \mathcal{H}$, the function $z \mapsto K(z, w) \cdot c$ belongs to $\mathcal{H}$ and

$$\langle f,\ K(\cdot, w)\,c \rangle_{\mathcal{H}} = \overline{c}^t f(w), \tag{6.4}$$

where $\langle \cdot, \cdot \rangle_{\mathcal{H}}$ is the scalar product in the Hilbert space $\mathcal{H}$; it is Eq. (6.4) that justifies the term "reproducing kernel"; it is proved that the function $K(z, w)$ is unique and is called the reproducing kernel of the Hilbert space $\mathcal{H}$; it is direct to prove that the function K is positive definite and is also Hermitian (in the usual complex setting), that is:

$$\overline{K(z, w)} = K(w, z).$$

We need to recall also that the reciprocal of the above is also true, that is, given a positive definite function $K(z, w) : \Omega \times \Omega \to \mathbb{C}^{n \times n}$ there exists a unique Hilbert space whose elements are functions from Ω to $\mathbb{C}^n$ and with reproducing kernel K.

An important example for the case $n = 1$ that we will use later is the function

$$K(z, w) := \frac{1}{1 - z\overline{w}} \quad z, w \in \mathbb{K}, \tag{6.5}$$

which is a positive definite function and it is known that its reproducing kernel Hilbert space is the Hardy space $\mathbf{H}^2(\mathbb{K})$.

In the previous theorem we have the particular case $n = 1$ and we have denoted by $\mathcal{H}(s)$ the reproducing kernel Hilbert space of functions holomorphic in $\mathbb{K}$ and with reproducing kernel $k_s(\lambda, \mu)$. This space provides a coisometric realization of s called the backward shift realization of s, which is already present (maybe in an implicit way) in the work of de Branges and Rovnyak [20]. See for instance Problem 47, p. 29 in that book.

Theorem 6.1.2 The operators

$$A[f](z) := \frac{f(z) - f(0)}{z},$$

$$(B[c])(z) := \frac{s(z) - s(0)}{z}c, \tag{6.6}$$

$$C[f] := f(0),$$

$$D[c] = s(0)c,$$

where $f \in \mathcal{H}(s)$ and $c \in \mathbb{C}$, define a coisometric realization of s.

Extending the notion of Schur function to the case of several complex variables is not so simple. We will focus on the cases of the polydisk and of the ball. We collect here in two theorems, [7], the analogs of Theorem 6.1.1 for the polydisk and the ball.

Theorem 6.1.3 Let s be holomorphic in the polydisk $\mathbb{K}^N$, let $\lambda = (\lambda_1, \ldots, \lambda_N) \in \mathbb{C}^N$ and let $\Lambda = \mathrm{diag}\,(\lambda_1, \lambda_2, \ldots, \lambda_N)$. Then, the following are equivalent:

(i) There exist $\mathbb{C}$–valued functions $k_1(\lambda, \mu), \ldots, k_N(\lambda, \mu)$, positive definite in $\mathbb{K}^N$ and such that

$$1 - s(\lambda)\,\overline{s(\mu)} = \sum_{n=1}^{N}(1 - \lambda_n\overline{\mu}_n)k_n(\lambda, \mu). \tag{6.7}$$

(ii) There exist Hilbert spaces $\mathcal{H}_1, \ldots, \mathcal{H}_N$ and a unitary operator matrix

$$\begin{pmatrix} A & B \\ C & D \end{pmatrix} : \mathcal{H} \oplus \mathbb{C} \longrightarrow \mathcal{H} \oplus \mathbb{C}$$

such that

$$s(\lambda) = D + C(I_{\mathcal{H}} - \Lambda A)^{-1}\Lambda B,$$

where $\mathcal{H} = \oplus_{n=1}^{N}\mathcal{H}_n$.

When $N = 1$, k_1 is uniquely determined, and is in fact the function k_s introduced in Theorem 6.1.1. Note that in general the decomposition (6.7) is not unique. Dividing both sides of (6.7) by $\prod_{n=1}^{N}(1 - \lambda_n \bar{\mu}_n)$ we see that in particular M_s is a contraction from the Hardy space of the polydisk $\mathbf{H}^2(\mathbb{K}^N)$ into itself. When $N > 2$ the class of Schur-Agler functions of the polydisk is strictly included in the class of functions holomorphic and contractive in the polydisk, or, equivalently, such that the multiplication operator $M_s[f] = sf$ is a contraction from the Hardy space $\mathbf{H}^2(\mathbb{K}^N)$ into itself. The two classes coincide when $N = 1$ or $N = 2$.

We now consider the case of the open unit ball $\mathbb{B}_N$ of $\mathbb{C}^N$. The function

$$\frac{1}{1 - \lambda \bar{\mu}} = \frac{1}{1 - \sum_{n=1}^{N} \lambda_n \bar{\mu}_n}$$

is positive-definite there. Its associated reproducing kernel Hilbert space is called the Drury-Arveson space, and will be denoted by $\mathcal{A}$. The space $\mathcal{A}$ is strictly included in the Hardy space $\mathbf{H}^2(\mathbb{B}_N)$ of the ball when $N > 1$, and the class of Schur-Agler functions of the ball is strictly included in the class of functions holomorphic and contractive in $\mathbb{B}_N$, or, equivalently, in the class of functions holomorphic in $\mathbb{B}_N$ and such that M_s is a contraction from $\mathbf{H}^2(\mathbb{B}_N)$ into itself.

Theorem 6.1.4 Let s be holomorphic in $\mathbb{B}_N$. The following are equivalent:

(i) The kernel

$$\frac{1 - s(\lambda)\,\overline{s(\mu)}}{1 - \lambda \bar{\mu}}$$

is positive definite in $\mathbb{B}_N$.

(ii) There exists a Hilbert space $\mathcal{H}$ and a unitary operator matrix

$$\begin{pmatrix} A & B \\ C & D \end{pmatrix} : \mathcal{H} \oplus \mathbb{C} \longrightarrow \mathcal{H} \oplus \mathbb{C}$$

such that

$$s(\lambda) = D + C(I_{\mathcal{H}} - \lambda A)^{-1}\lambda B.$$

(iii) The operator M_s is a contraction from the Drury-Arveson space into itself.

6.2 The Bicomplex Hardy Space

We now want to show how the ideas and concepts of Sect. 6.1 can be translated to the case of bicomplex analysis and bicomplex holomorphic functions. Let

$$\Omega_{\mathbb{K}^2} := \left\{ Z = z_1 + z_2\mathbf{j} = \mathbf{e}\,\beta_1 + \mathbf{e}^\dagger \beta_2 \mid (\beta_1, \beta_2) \in \mathbb{K}^2 \right\}, \tag{6.8}$$

with $\mathbb{K}$, as in the previous section, being the unit disk in the complex plane and $\mathbb{K}^2 = \mathbb{K} \times \mathbb{K}$. The reader may consider redundant the notation $\Omega_{\mathbb{K}^2}$, but recall that there are many different ways to identify the set $\mathbb{BC}$ with $\mathbb{C}^2$ and we want to emphasize that now we are making this identification via the idempotent representation.

We define the bicomplex Hardy space $\mathbf{H}^2(\Omega_{\mathbb{K}^2})$ to be the set of functions $f :$ $\Omega_{\mathbb{K}^2} \to \mathbb{BC}$ such that for any $Z \in \Omega_{\mathbb{K}^2}$:

$$f(Z) = \sum_{n=0}^{\infty} f_n \cdot Z^n,$$

where $\forall n \in \mathbb{N}$, $f_n \in \mathbb{BC}$ and the series of hyperbolic numbers $\sum_{n=0}^{\infty} |f_n|_{\mathbf{k}}^2$ is convergent. Setting $f_n = \mathbf{e} f_{n1} + \mathbf{e}^\dagger f_{n2}$ one gets:

$$|f_n|_{\mathbf{k}}^2 = \mathbf{e}|f_{n1}|^2 + \mathbf{e}^\dagger |f_{n2}|^2,$$

which means that both series

$$\sum_{n=0}^{\infty} |f_{n1}|^2, \qquad \sum_{n=0}^{\infty} |f_{n2}|^2$$

are convergent and thus both functions

$$f_1(\beta_1) := \sum_{n=0}^{\infty} f_{n1}\beta_1^n, \qquad f_2(\beta_2) := \sum_{n=0}^{\infty} f_{n2}\beta_2^n$$

belong to the Hardy space of the unit disk $\mathbf{H}^2(\mathbb{K})$; of course

$$f(Z) = \mathbf{e} \cdot f_1(\beta_1) + \mathbf{e}^\dagger \cdot f_2(\beta_2),$$

or, equivalently,

$$f(z_1 + \mathbf{j}z_2) = \mathbf{e} \cdot f_1(z_1 - iz_2) + \mathbf{e}^\dagger \cdot f_2(z_1 + iz_2).$$

This means that the bicomplex Hardy space can be written as

$$\mathbf{H}^2(\Omega_{\mathbb{K}^2}) = \mathbf{e} \cdot \mathbf{H}^2(\mathbb{K}) + \mathbf{e}^\dagger \cdot \mathbf{H}^2(\mathbb{K}). \tag{6.9}$$

The $\mathbb{BC}$-valued inner product on $\mathbf{H}^2(\Omega_{\mathbb{K}^2})$ is given by

$$\langle f, g \rangle_{\mathbf{H}^2(\Omega_{\mathbb{K}^2})} := \sum_{n=0}^{\infty} f_n \cdot g_n^*$$

$$= \mathbf{e} \cdot \sum_{n=0}^{\infty} f_{n1} \cdot \overline{g_{n1}} + \mathbf{e}^\dagger \cdot \sum_{n=0}^{\infty} f_{n2} \cdot \overline{g_{n2}}$$

$$= \mathbf{e} \cdot \langle f_1, g_1 \rangle_{\mathbf{H}^2(\mathbb{K})} + \mathbf{e}^\dagger \cdot \langle f_2, g_2 \rangle_{\mathbf{H}^2(\mathbb{K})},$$

and it generates the $\mathbb{D}^+$-valued norm:

$$\| f \|^2_{\mathbb{D}, \mathbf{H}^2(\Omega_{\mathbb{K}^2})} = \langle f, f \rangle_{\mathbf{H}2(\Omega_{\mathbb{K}^2})} = \sum_{n=0}^{\infty} |f_n|^2_{\mathbf{k}}$$

$$= \mathbf{e} \cdot \sum_{n=0}^{\infty} |f_{n1}|^2 + \mathbf{e}^\dagger \cdot \sum_{n=0}^{\infty} |f_{n2}|^2$$

$$= \mathbf{e} \cdot \| f_1 \|^2_{\mathbf{H}^2(\mathbb{K})} + \mathbf{e}^\dagger \cdot \| f_2 \|^2_{\mathbf{H}^2(\mathbb{K})}.$$

Thus one can say that we have obtained an analog of the classic Hardy space on the unit disk for the bicomplex setting where we deal with the bidisk, but now in the idempotent, not cartesian, coordinates; the inner product in this Hardy space is $\mathbb{BC}$-valued and (what is even more remarkable) the "adequate" norm takes values in $\mathbb{D}^+$, the positive hyperbolic numbers.

6.3 Positive Definite Functions

Let Ω be some set. The $\mathbb{BC}^{n \times n}$-valued function $K(z, w)$ defined for $z, w \in \Omega$ is said to be positive definite if it is $*$–Hermitian:

$$K(z, w) = K(w, z)^{*t}, \quad \forall z, w \in \Omega,$$

and if for every choice of $N \in \mathbb{N}$, of columns $c_1, \dots, c_N \in \mathbb{C}^n(\mathbf{i})$ and of $z_1, \dots, z_N \in \Omega$ one has that

$$\sum_{\ell, j=1}^{N} c_j^{*t} K(z_j, z_\ell) c_\ell \in \mathbb{D}^+.$$

Proposition 6.3.1 The $\mathbb{BC}^{n \times n}$-valued function

$$K(z, w) = K_1(z, w) + \mathbf{j} K_2(z, w)$$

is positive definite if and only if $K_1(z, w)$ is positive definite and $\mathbf{i} K_2(z, w)$ is Hermitian and such that

$$- K_1(z, w) \leq \mathbf{i} K_2(z, w) \leq K_1(z, w). \tag{6.10}$$

Proof:

This follows from (2.10) applied to the matrices

$$\begin{pmatrix} K(z_1, z_1) & K(z_1, z_2) & \cdots & K(z_1, z_N) \\ K(z_2, z_1) & K(z_2, z_2) & \cdots & K(z_2, z_N) \\ \vdots & \vdots & & \vdots \\ K(z_N, z_1) & K(z_N, z_2) & \cdots & K(z_N, z_N) \end{pmatrix}.$$

$\square$

Corollary 6.3.2 Let $K(z, w)$ be $\mathbb{BC}^{n \times n}$-valued function defined in Ω^2. Write $K(z, w) = k_1(z, w)\mathbf{e} + k_2(z, w)\mathbf{e}^\dagger$, where k_1 and k_2 are $\mathbb{C}^{n \times n}$-valued functions. Then, K is positive-definite in Ω if and only if the functions k_1 and k_2 are positive definite in Ω.

Theorem 6.3.3 Let K be a $\mathbb{BC}^{n \times n}$-valued function positive definite on the set Ω. There exists a unique reproducing kernel Hilbert space $\mathscr{H}(K)$ of functions defined on Ω with reproducing kernel K. Namely, for all $w \in \Omega, c \in \mathbb{BC}^n$ and $f \in \mathscr{H}(K)$ one has that

(i) The function $z \mapsto K(z, w)c$ belongs to $\mathscr{H}(K)$, and
(ii) The reproducing kernel property

$$\langle f, K(\cdot, w)c \rangle_{\mathscr{H}(K)} = c^{*t} f(w)$$

holds.

Proof:

Given $K = K_1 + \mathbf{j}K_2 = \mathbf{e} \cdot (K_1 - \mathbf{i}K_2) + \mathbf{e}^\dagger \cdot (K_1 + \mathbf{i}K_2)$, a $\mathbb{BC}^{n \times n}$-valued positive definite function on a set Ω, the associated reproducing kernel Hilbert space consists of the functions of the form

$$F(z) = f_1(z)\,\mathbf{e} + f_2(z)\,\mathbf{e}^\dagger,$$

where $f_1 \in \mathscr{H}(K_1 - \mathbf{i}K_2)$ and $f_2 \in \mathscr{H}(K_1 + \mathbf{i}K_2)$, and with $\mathbb{BC}$-valued inner product

$$\langle F, G\rangle_{\mathscr{H}(K)} := \mathbf{e} \langle f_1, g_1\rangle_{\mathscr{H}(K_1 - iK_2)} + \mathbf{e}^\dagger \langle f_2, g_2\rangle_{\mathscr{H}(K_1 + iK_2)}. \tag{6.11}$$

$\square$

In the previous theorem, the $\mathbb{C}^{n \times n}$-valued kernels are in particular positive definite in Ω. We denote by $\mathscr{H}(K_1 - iK_2)$ and $\mathscr{H}(K_1 + iK_2)$ the associated reproducing kernel Hilbert spaces of $\mathbb{C}^n$-valued functions on Ω.

Corollary 6.3.4 The function K is positive definite in Ω if and only if there is a $\mathbb{B}\mathbb{C}$-Hilbert space $\mathcal{H}$ and a function f from Ω into $\mathcal{H}$ such that

$$K(z, w) = \langle f(w), f(z)\rangle_{\mathcal{H}} \quad \forall z, w \in \Omega. \tag{6.12}$$

Proof:
Write $K(z, w) = k_1(z, w)\mathbf{e} + k_2(z, w)\mathbf{e}^\dagger$, where k_1 and k_2 are complex-valued positive definite functions. Then there exist complex Hilbert spaces $\mathcal{H}_1$ and $\mathcal{H}_2$ and functions f_1 and f_2, defined on Ω and with values in $\mathcal{H}_1$ and $\mathcal{H}_2$, respectively, such that for any $z, w \in \Omega$ there holds:

$$k_1(z, w) = \langle f_1(w), f_1(z)\rangle_{\mathcal{H}_1} \quad \text{and} \quad k_2(z, w) = \langle f_2(w), f_2(z)\rangle_{\mathcal{H}_2}.$$

Then using the process described in Sect. 3.2, define $\mathcal{H}$ to be

$$\mathcal{H} = \mathbf{e} \cdot \mathcal{H}_1 + \mathbf{e}^\dagger \cdot \mathcal{H}_2.$$

Applying to this $\mathbb{B}\mathbb{C}$–module the results from Sect. 4.3.3 and in particular the formula (4.6), the inner product on $\mathcal{H}$ is given by

$$\begin{aligned} \langle h, g\rangle_{\mathcal{H}} &= \langle \mathbf{e}\, h_1 + \mathbf{e}^\dagger\, h_2, \mathbf{e}\, g_1 + \mathbf{e}^\dagger\, g_2\rangle_{\mathcal{H}} \\ &= \mathbf{e} \langle h_1, g_1\rangle_{\mathcal{H}_1} + \mathbf{e}^\dagger \langle h_2, g_2\rangle_{\mathcal{H}_2}, \end{aligned}$$

and for any $z \in \Omega$:

$$f(z) := \mathbf{e} \cdot f_1(z) + \mathbf{e}^\dagger \cdot f_2(z);$$

thus one has:

$$\begin{aligned} \langle f(w), f(z)\rangle_{\mathcal{H}} &= \langle \mathbf{e}\, f_1(w) + \mathbf{e}^\dagger\, f_2(w), \mathbf{e}\, f_1(z) + \mathbf{e}^\dagger\, f_2(z)\rangle_{\mathcal{H}} \\ &= \mathbf{e} \langle f_1(w), f_1(z)\rangle_{\mathcal{H}_1} + \mathbf{e}^\dagger \langle f_2(w), f_2(z)\rangle_{\mathcal{H}_2} \\ &= \mathbf{e}\, k_1(z, w) + \mathbf{e}^\dagger\, k_2(z, w) = K(z, w). \end{aligned}$$

This concludes the proof. $\square$

In the complex case, there is a one-to-one correspondence between positive definite functions on a set Ω and reproducing kernel Hilbert spaces of functions defined on Ω. In the present case, one has the following result:

Theorem 6.3.5 There is a one-to-one correspondence between $\mathbb{BC}^{n\times n}$-valued positive definite functions on a set Ω and reproducing kernel Hilbert spaces of $\mathbb{BC}^n$-valued functions defined on Ω.

We conclude this section with an example.

Theorem 6.3.6

(1) The function $\dfrac{1}{1 - ZW^*}$ is positive definite in $\Omega_{\mathbb{K}^2}$.

(2) Its associated $\mathbb{BC}$ reproducing kernel Hilbert space is $\mathbf{H}^2(\Omega_{\mathbb{K}^2})$.

(3) For every $f \in \mathbf{H}^2(\Omega_{\mathbb{K}^2})$, and every $a \in \Omega_{\mathbb{K}^2}$, we have:

$$\langle f,\, (1 - Za^*)^{-1}\rangle_{\mathbf{H}^2(\Omega_{\mathbb{K}^2})} = f(a).$$

Proof:

(1) Writing $Z = \beta_1\mathbf{e} + \mu_1\mathbf{e}^\dagger$ and $W = \beta_2\mathbf{e} + \mu_2\mathbf{e}^\dagger$, where $\beta_1, \beta_2, \mu_1, \mu_2$ are in $\mathbb{K}$, we have:

$$\frac{1}{1 - ZW^*} = \frac{1}{1 - \beta_1\overline{\mu}_1}\mathbf{e} + \frac{1}{1 - \beta_2\overline{\mu}_2}\mathbf{e}^\dagger.$$

The claim follows then from Corollary 6.3.2.

Items (2) and (3) are consequences of Corollary 6.3.3, and of (6.9). □

6.4 Schur Multipliers and Their Characterizations in the $\mathbb{BC}$-Case

Definition 6.4.1 A $\mathbb{BC}^{n\times m}$-valued $\mathbb{BC}$-holomorphic function s on $\Omega_{\mathbb{K}^2}$ is called a Schur function if it can be written as

$$s(Z) = s_1(\beta_1)\mathbf{e} + s_2(\beta_2)\mathbf{e}^\dagger,$$

where s_1 and s_2 belong to $\mathscr{S}^{n\times m}(\mathbb{K})$.

We will use the notation $\mathscr{S}^{n\times m}(\Omega_{\mathbb{K}^2})$ for these Schur functions s.

Theorem 6.4.2 Let s be a $\mathbb{BC}^{n\times m}$-valued $\mathbb{BC}$-holomorphic function on $\Omega_{\mathbb{K}^2}$. Then, the following are equivalent:

(1) s belongs to $\mathscr{S}^{n\times m}(\Omega_{\mathbb{K}^2})$.

(2) $s(Z)s(Z)^{*t} \leq I_n, \quad \forall Z \in \Omega_{\mathbb{K}^2}$.

(3) The function

$$\frac{I_n - s(Z)s(W)^{*t}}{1 - ZW^*} \tag{6.13}$$

is positive definite in $\Omega_{\mathbb{K}^2}$.

(4) The operator of multiplication by s is a $\mathbb{D}$–contraction from $(\mathbf{H}^2(\Omega_{\mathbb{K}^2}))^m$ into $(\mathbf{H}^2(\Omega_{\mathbb{K}^2}))^n$, meaning that

$$\langle sf, sf \rangle_{(\mathbf{H}^2(\Omega_{\mathbb{K}^2}))^n} \leq \langle f, f \rangle_{(\mathbf{H}^2(\Omega_{\mathbb{K}^2}))^m}, \quad \forall f \in (\mathbf{H}^2(\Omega_{\mathbb{K}^2}))^m.$$

(5) s admits a realization

$$s(Z) = D + ZC(I_{\mathcal{H}} - ZA)^{-1}B \tag{6.14}$$

where $\mathcal{H}$ is a $\mathbb{BC}$-Hilbert space and the operator matrix

$$\begin{pmatrix} A & B \\ C & D \end{pmatrix} : \mathcal{H} \oplus \mathbb{BC}^m \longrightarrow \mathcal{H} \oplus \mathbb{BC}^n$$

is coisometric.

Proof:

Assume that (1) holds. Then, with $W = \gamma_1 \mathbf{e} + \gamma_2 \mathbf{e}^{\dagger}$, we can write

$$I_n - s(Z)s(W)^{*t} = (I_n - s_1(\beta_1)s_1(\gamma_1)^*)\mathbf{e} + (I_n - s_2(\beta_2)s_2(\gamma_2)^*)\mathbf{e}^{\dagger}.$$

Since s_1 and s_2 are classical Schur functions we have:

$$I_n - s_1(\beta_1)s_1(\beta_1)^* \geq 0 \quad \text{and} \quad I_n - s_2(\beta_2)s_2(\beta_2)^* \geq 0$$

for all $\beta_1, \beta_2 \in \mathbb{K}$. From Proposition 2.2.7 we get that (2) holds. When (2) holds both the kernels

$$\frac{I_n - s_1(\beta_1)\overline{s_1(\gamma_1)}^t}{1 - \beta_1\overline{\gamma_1}} \quad \text{and} \quad \frac{I_n - s_2(\beta_2)\overline{s_2(\gamma_2)}^t}{1 - \beta_2\overline{\gamma_2}} \tag{6.15}$$

are positive definite in the open unit disk. Writing

$$\frac{I_n - s(Z)s(W)^{*t}}{1 - ZW^*} = \frac{I_n - s_1(\beta_1)\overline{s_1(\gamma_1)}^t}{1 - \beta_1\overline{\gamma_1}}\mathbf{e} + \frac{I_n - s_2(\beta_2)\overline{s_2(\gamma_2)}^t}{1 - \beta_2\overline{\gamma_2}}\mathbf{e}^{\dagger},$$

it follows from Corollary 6.3.2 that the kernel (6.13) is positive definite in $\Omega_{\mathbb{K}^2}$. Thus (3) holds. When (3) holds, the kernels (6.15) are in particular positive definite in the open unit disk. Thus the operators of multiplication by s_1 and s_2 are contractions from $(\mathbf{H}^2(\mathbb{K}))^m$ into $(\mathbf{H}^2(\mathbb{K}))^n$, that is,

$$\langle s_j f_j, s_j f_j \rangle_{(\mathbf{H}^2(\mathbb{K}))^n} \le \langle f_j, f_j \rangle_{(\mathbf{H}^2(\mathbb{K}))^m}, \quad j = 1, 2.$$

Thus

$$\begin{aligned}
\langle sf, sf \rangle_{(\mathbf{H}^2(\Omega_{\mathbb{K}^2}))^n} &= \langle s_1 f_1, s_1 f_1 \rangle_{(\mathbf{H}^2(\mathbb{K}))^n} \mathbf{e} + \langle s_2 f_2, s_2 f_2 \rangle_{(\mathbf{H}^2(\mathbb{K}))^n} \mathbf{e}^\dagger \\
&\le \langle f_1, f_1 \rangle_{(\mathbf{H}^2(\mathbb{K}))^m} \mathbf{e} + \langle f_2, f_2 \rangle_{(\mathbf{H}^2(\mathbb{K}))^m} \mathbf{e}^\dagger \\
&= \langle f, f \rangle_{(\mathbf{H}^2(\Omega_{\mathbb{K}^2}))^m},
\end{aligned}$$

that is, (4) holds. Assume now that (4) holds. By definition of the inequality, s_1 and s_2 are classical Schur functions. Using Theorem 6.1.1 we can write

$$s_j(\beta_j) = D_j + \beta_j C_j (I_{\mathscr{H}_j} - \beta_j A_j)^{-1} B_j, \quad j = 1, 2,$$

where in these expressions $\mathscr{H}_1$ and $\mathscr{H}_2$ are complex Hilbert spaces and the operator matrices

$$\begin{pmatrix} A_j & B_j \\ C_j & D_j \end{pmatrix} : \mathcal{H}_j \oplus \mathbb{C}^m \longrightarrow \mathcal{H}_j \oplus \mathbb{C}^n$$

are coisometric. Realization (6.14) follows by taking into account again the process described in Sect. 3.2 and setting

$$\mathcal{H} := \mathbf{e} \cdot \mathcal{H}_1 + \mathbf{e}^\dagger \mathcal{H}_2.$$

Then the operators

$$\begin{aligned}
A &:= \mathbf{e}\, A_1 + \mathbf{e}^\dagger A_2, \\
B &:= \mathbf{e}\, B_1 + \mathbf{e}^\dagger B_2, \\
C &:= \mathbf{e}\, C_1 + \mathbf{e}^\dagger C_2, \\
D &:= \mathbf{e}\, D_1 + \mathbf{e}^\dagger D_2,
\end{aligned}$$

are such that

$$s(Z) = D + ZC(I_{\mathcal{H}} - ZA)^{-1} B$$

and they form a coisometric matrix operator, that is, (5) holds. When (5) holds we have for every $Z, W \in \Omega_{\mathbb{K}^2}$:

$$\frac{I_n - s(Z)s(W)^{*t}}{1 - ZW^*} = C(I_{\mathcal{H}} - ZA)^{-1}(I_{\mathcal{H}} - WA)^{-1} C^\sharp.$$

Setting $Z = W$ we see that (2) holds, which implies (1) as one can see by considering the idempotent components. $\qquad\square$

6.5 An Example: Bicomplex Blaschke Factors

We conclude this work by presenting an example of bicomplex Schur multiplier, namely, a bicomplex Blaschke factor. Let $a = a_1 + \mathbf{j}\, a_2 \in \Omega_{\mathbb{K}^2}$. We set

$$b_a(Z) := \frac{Z - a}{1 - Za^*}. \tag{6.16}$$

Clearly, the function b_a is defined for all Z such that

$$(z_1 - \mathbf{i}z_2)(\overline{a_1} + \mathbf{i}\,\overline{a_2}) \neq 1, \quad \text{and} \quad (z_1 + \mathbf{i}z_2)(\overline{a_1} - \mathbf{i}\,\overline{a_2}) \neq 1, \tag{6.17}$$

and in particular for all $Z \in \Omega_{\mathbb{K}^2}$. We will call b_a the bicomplex Blaschke factor with zero a when $aa^* \neq 1$. As in the complex case,

$$b_a(Z) = -a + Z\frac{1 - aa^*}{1 - Za^*},$$

so that a realization of $b_a(Z)$ is given by

$$\begin{pmatrix} a^* & 1 \\ 1 - aa^* & -a \end{pmatrix}.$$

This realization is not coisometric, but is readily fixed as follows. Let $a = \eta_1 \mathbf{e} + \eta_2 \mathbf{e}^\dagger$, and

$$u = \sqrt{1 - |\eta_1|^2}\ \mathbf{e} + \sqrt{1 - |\eta_2|^2}\ \mathbf{e}^\dagger.$$

Then $u \in \mathbb{D}^+$ and $u^2 = 1 - aa^*$. The matrix

$$\begin{pmatrix} A & B \\ C & D \end{pmatrix} = \begin{pmatrix} a^* & u \\ u & -a \end{pmatrix}$$

also defines a realization of s, and it is unitary (and in particular co-isometric).

Proposition 6.5.1 It holds that

$$b_a(Z)b_a(Z)^* = \begin{cases} \ll 1, & \text{for } Z \in \Omega_{\mathbb{K}^2}, \\ \\ 1, & \text{for } ZZ^* = 1. \end{cases} \tag{6.18}$$

Proof:

Let $Z \in \Omega_{\mathbb{K}^2}$. Then $1 - Za^*$ is invertible and we have:

$$
\begin{aligned}
1 - b_a(Z)b_a(Z)^* &= 1 - \frac{(Z-a)(Z-a)^*}{(1-Za^*)(1-Z^*a)} \\
&= \frac{(1-Za^*)(1-Z^*a) - (Z-a)(Z-a)^*}{(1-Za^*)(1-Z^*a)} \\
&= \frac{(1-ZZ^*)(1-aa^*)}{(1-Za^*)(1-Z^*a)},
\end{aligned}
$$

which belongs to $\mathbb{D}^+$ as a product and quotient of positive hyperbolic numbers.

Let now $ZZ^* = 1$. The element Z is in particular invertible and we can write for such Z

$$
b_a(Z) = \frac{Z-a}{Z(Z^*-a^*)}.
$$

Thus, for such Z,

$$
b_a(Z)b_a(Z)^* = \frac{(Z-a)(Z-a)^*}{ZZ^*(Z^*-a^*)(Z-a)} = 1.
$$

$\square$

We note that both

$$
b_{a,1}(\beta_1) = \frac{\beta_1 - \eta_1}{1 - \beta_1 \overline{\eta_1}} \quad \text{and} \quad b_{a,2}(\beta_2) = \frac{\beta_2 - \eta_2}{1 - \beta_2 \overline{\eta_2}} \tag{6.19}
$$

are classical Blaschke factors.

We will study in a future publication interpolation problems in the present setting. Here we only mention the following result:

Proposition 6.5.2 Let $a \in \Omega_{\mathbb{K}^2}$. Then:

(1) For every $f \in \mathbf{H}^2(\Omega_{\mathbb{K}^2})$ it holds that

$$
\langle b_a f, b_a f \rangle_{\mathbf{H}^2(\Omega_{\mathbb{K}^2})} = \langle f, f \rangle_{\mathbf{H}^2(\Omega_{\mathbb{K}^2})},
$$

that is, the operator of multiplication by b_a is $\mathbb{D}$-isometric.

(2) A function f belongs to $\mathbf{H}^2(\Omega_{\mathbb{K}^2})$ and vanishes at the point a if and only if it can be written as $f = b_a g$ with $g \in \mathbf{H}^2(\Omega_{\mathbb{K}^2})$. In this case,

$$
\langle f, f \rangle_{\mathbf{H}^2(\Omega_{\mathbb{K}^2})} = \langle g, g \rangle_{\mathbf{H}^2(\Omega_{\mathbb{K}^2})}.
$$

i.e., their $\mathbb{D}$-norms coincide.

Proof:

(1) We first note that for $Z \in \Omega_{\mathbb{K}^2}$ we have:

$$b_a(Z) = -a + (1 - aa^*) \sum_{n=1}^{\infty} Z^n (a^*)^{n-1},$$

where the convergence is in $\mathbf{H}^2(\Omega_{\mathbb{K}^2})$, and also pointwise, since we are in a reproducing kernel Hilbert space. Thus, for $N \in \mathbb{N}$ we have:

$$b_a(Z)Z^N = -aZ^N + (1 - aa^*) \sum_{n=1}^{\infty} Z^{n+N} (a^*)^{n-1}. \qquad (6.20)$$

Denote $\mathbb{N}_0 := \mathbb{N} \cup \{0\}$. Thus, for $N \in \mathbb{N}_0$,

$$
\begin{aligned}
\langle b_a(Z)Z^N, b_a(Z)Z^N \rangle_{\mathbf{H}^2(\Omega_{\mathbb{K}^2})} &= aa^* + (1 - aa^*)^2 \sum_{n=1}^{\infty} (aa^*)^{n-1} \\
&= aa^* + (1 - aa^*)^2 (1 - aa^*)^{-1} \\
&= 1.
\end{aligned}
$$

Let now $N, M \in \mathbb{N}_0$ with $N < M$, and let $M = N + H$. We have in view of (6.20):

$$
\begin{aligned}
\langle b_a(Z)Z^N, b_a(Z)Z^M \rangle_{\mathbf{H}^2(\Omega_{\mathbb{K}^2})} &= -a^*(1 - aa^*)(a^*)^{H-1} \\
&\quad + (1 - aa^*) \sum_{n=0}^{\infty} (a^*)^{H+n-1} a^{n-1} \\
&= -a^*(1 - aa^*)(a^*)^{H-1} \\
&\quad + (1 - aa^*)^2 (a^*)^H (1 - aa^*)^{-1} \\
&= 0.
\end{aligned}
$$

Hence, for any polynomial p we have

$$\langle b_a p, b_a p \rangle_{\mathbf{H}^2(\Omega_{\mathbb{K}^2})} = \langle p, p \rangle_{\mathbf{H}^2(\Omega_{\mathbb{K}^2})},$$

and the result follows by continuity for every element in $\mathbf{H}^2(\Omega_{\mathbb{K}^2})$.

(2) One direction is clear. If $f = b_a g$ with $g \in \mathbf{H}^2(\Omega_{\mathbb{K}^2})$, then $f \in \mathbf{H}^2(\Omega_{\mathbb{K}^2})$ (this can be verified component-wise) and vanishes at the point a. Conversely, assume that $f(a) = 0$, and let $a = a_1 + \mathbf{j} a_2$. Then $f_1(\eta_1) = f_2(\eta_2) = 0$, where

$$\eta_1 = a_1 - \mathbf{i} a_2 \quad \text{and} \quad \eta_2 = a_1 + \mathbf{i} a_2.$$

By the classical theory we have (with $b_{a,1}$ and $b_{a,2}$ defined by (6.19)):

$$f_j = b_{a,j} g_j, \quad j = 1, 2,$$

with $g_j \in \mathbf{H}^2(\mathbb{K})$. The result follows by regrouping the components. $\qquad\square$

As we have noted already in the Introduction the classic, complex Schur analysis has connections and applications to many interesting problems, thus we expect that our brief study of some basic facts of bicomplex Schur analysis will become a starting point for deeper developments of it together with many applications.

References

1. D. Alpay, *The Schur algorithm, reproducing kernel spaces and system theory* Translated from the 1998 French original by Stephen S. Wilson Panoramas et Synthèses. (American Mathematical Society, 2001)
2. I. Schur, Über die Potenzreihen, die im Innern des Einheitkreises beschränkten sind, I. J. Reine Angew. Math 147 (1917), 205–232. English translation. In: I. Schur methods in operator theory and signal processing. (Operator Theory: Advances and Applications OT 18 (1986). (Birkhäuser Verlag, Basel, 1986)
3. I. Schur, Über die Potenzreihen, die im Innern des Einheitkreises Beschränkt sind, II. J. Reine Angew. Math 148 (1918), 122–145. English translation. In: I. Schur methods in operator theory and signal processing. (Operator Theory: Advances and Applications OT 18 (1986). (Birkhäuser Verlag, Basel, 1986)
4. C. Carathéodory, L. Fejér, Uner den Zusammenhang der Extremen von Harmonischen Funktionen mit ihren Koeffzienten und uber den Picard-Landauschen Satz. Rend. Circ. Mat. Palermo **32**(2), 218–251 (1911)
5. G. Herglotz, Uber Potenzenreihen mit positiven reelle Teil im Einheitskreis. Sitzungsber Sachs. Akad. Wiss. Leipzig, Math **63**, 501–511 (1911)
6. O. Toeplitz, Uber die Fouriersche Entwicklung positiver Funktionen. Rend. Circ. Mat. Palermo **32**(2), 191–192 (1911)
7. J. Agler, On the representation of certain holomorphic functions defined on a polydisk. Oper. Theor. Adv. Appl. **48**, 47–66 (1990)
8. J. Agler, J. McCarthy, Complete Nevanlinna-Pick kernels. JFUNA2, **175**, 111–124 (2000)
9. D. Alpay, P. Dewilde, H. Dym, Lossless inverse scattering and reproducing kernels for upper triangular operators. Extension and interpolation of linear operators and matrix, functions, Birkhäuser, Basel. Oper. Theory Adv. Appl. **47**, 61–135 (1990)
10. D. Alpay, D. Levanony, Rational functions associated with the white noise space and related topics. Potential Anal. **29**, 195–220 (2008)
11. D. Alpay, V. Vinnikov, Finite dimensional de Branges spaces on Riemann surfaces. J. Funct. Anal. **189**(2), 283–324 (2002)
12. D. Alpay, M. Mboup, A characterization of Schur multipliers between character-automorphic Hardy spaces. Integr. Eqn. Oper. Theory **62**, 455–463 (2008)
13. D. Alpay, M. Mboup, Discrete-time multi-scale systems. Integr. Eqn. Oper. Theory **68**, 163–191 (2010)
14. D. Alpay, M. Shapiro, D. Volok, Espaces de de Branges Rovnyak: le cas hyper-analytique. Comptes Rendus Mathématiques **338**, 437–442 (2004)
15. D. Alpay, M. Shapiro, D. Volok, Rational hyperholomorphic functions in R^4. J. Funct. Anal. **221**(1), 122–149 (2005)
16. D. Alpay, M. Shapiro, D. Volok, Reproducing kernel spaces of series of Fueter polynomials. Operator theory in Krein spaces and nonlinear eigenvalue problems. Oper. Theory Adv. Appl. **162**, 19–45 (2006)

17. D. Alpay, F.M. Correa-Romero, M.E. Luna-Elizarrarás, M. Shapiro, Hyperholomorphic rational functions: the Clifford analysis case. Complex Var. Elliptic Equ. An Int. J. **52**(1), 59–78 (2007)
18. D. Alpay, F. Colombo, I. Sabadini, Schur functions and their realizations in the slice hyperholomorphic setting. Integr. Eqn. Oper. Theory **72**, 253–289 (2012)
19. W. Rudin, *Analyse réelle et Complexe* (Masson, Paris, 1980)
20. L. de Branges, J. Rovnyak, *Square Summable Power Series.* (Holt, Rinehart and Winston, New York , 1966)

Printed by Publishers' Graphics LLC
ASO140406.20.06.4